The Open University

BLOCK 4

EVOLUTION OF CONTINENTAL CRUST

Prepared for the Course Team by

NIGEL HARRIS

THE S267 COURSE TEAM

CHAIRMAN

Peter J Smith

C J Hawkesworth

COURSE COORDINATOR

Veronica M E Barnes

COURSE MANAGER

Val Russell

AUTHORS

Andrew Bell

Stephen Blake

Nigel Harris

David A Rothery

Hazel Rymer

EDITORS

David Tillotson

Gerry Bearman

Sue Glover

DESIGNER

Caroline Husher

GRAPHIC ARTIST

Alison George

BBC

David Jackson

The Open University, Walton Hall, Milton Keynes, MK7 6AA.

First published 1993.

Edited, designed and typeset in the United Kingdom by the Open University.

Printed in the United Kingdom by Thanet Press Limited, Margate, Kent

ISBN 0 7492 8165 0

This text forms part of an Open University Second Level Course. If you would like a copy of *Studying with the Open University*, please write to the Central Enquiry Service, PO Box 200, The Open University, Walton Hall, Milton Keynes, MK7 6YZ. If you have not already enrolled on the Course and would like to buy this or other Open University material, please write to Open University Educational Enterprises Ltd, 12 Cofferidge Close, Stony Stratford, Milton Keynes, MK11 1BY, United Kingdom.

3.1

S267b4i3.1

S267
HOW THE EARTH WORKS:
The Earth's Interior

BLOCK 4
EVOLUTION OF CONTINENTAL CRUST

4.1 INTRODUCTION AND STUDY GUIDE

Block 4 describes how and where the continental crust forms. This is done by considering the mineralogy and chemistry of magmas formed at island arcs, at active continental margins and within collision zones. The three main Sections examine each of these tectonic settings in turn, building directly on what you have learned about plate boundaries in Block 2 and on your geochemical skills developed in Block 3. In this Block, you will be looking at progressively more silicic magmas compared to those that you have examined in Block 3. Two new ternary systems are introduced (diopside–anorthite–albite in Section 4.2.4 and quartz–orthoclase–albite in Section 4.3.2) which will integrate binary solid solutions with ternary plots. The video 'Melts to Minerals' (VB 04) forms a useful revision of what you have learned about phase diagrams, and introduces you to the new systems. To help with the challenge of working with more complex diagrams, we provide models of the new ternary systems in your Kit which are described in AV 08 ('Phase Diagrams').

Two new techniques are introduced in the Block. Firstly, graphical and numerical skills are required for interpreting naturally occurring radioactive isotopes (Section 4.3.3). Isotopes are a vital tool for the geochemist, not only for determining the age of an igneous rock, but also for tracing the geochemical characteristics of its source region. In AV 09 ('Isotope Geochemistry and Geochronology'), the details of Rb–Sr isotopes and their applications are explained step-by-step. The second new area of study is the use of minerals from metamorphic rocks as thermometers and barometers. An introduction to metamorphic reactions (Section 4.4.1) provides a basis for understanding the processes that occur in zones of continent–continent collision which culminate in magma formation.

Major elements, trace elements and isotopes provide different lines of evidence for interpreting the different stages of magma formation and of crustal evolution. These are brought together in 'Magmas in Scotland' (VB 06), which examines the origin of basalts and granites from the Isle of Skye, and should be viewed towards the end of Section 4.3.

The continental crust comprises a wider range of minerals and a wider range of rock types than does the oceanic crust. Hand specimens provided in the Kit (described in AV 07 'Home Kit Rocks'), and outcrops illustrated in the Colour Plate Section, are representative of the common igneous and metamorphic rocks that characterize the continental crust. You will also find useful the two appendixes at the end of Block 3, which summarize the important features of common rocks and minerals. Although you are not required to memorize details of either the silicate minerals (Appendix 1) or the composition of the different igneous rocks (Appendix 2), we have collated this information in the appendixes for you to refer to as the various rocks and minerals are introduced.

We estimate the study time for the different Sections in terms of Course-Unit Equivalents to be:

Section 4.2 – 1.25 CUE

Section 4.3 – 1.25 CUE

Section 4.4 – 0.5 CUE

In this Block, we turn from the oceanic crust to the continental crust. Oceanic crust is made up largely of basalts and gabbros. In contrast, the continents provide a wider range of igneous rock compositions that are much more silicic. These extend from basaltic compositions ($SiO_2 \approx 50\%$) to granites with an SiO_2 content of more than 70% (Table 4.1). The variation in silica contents is accompanied by other chemical changes together with significant physical differences. Silicic magmas have a lower density and a lower melting point than basaltic magmas. All these differences have important consequences for understanding how the continents evolve.

Table 4.1 Typical chemical compositions (in mass % of oxides) for some common igneous rocks (plutonic rock types in capital letters, volcanic rocks of the same chemical composition in lower case).

	GRANITE	GRANODIORITE	DIORITE	GABBRO	PERIDOTITE
	rhyolite	dacite	andesite	basalt	
SiO_2	72.1	66.9	53.9	48.4	43.5
TiO_2	0.37	0.57	1.5	1.3	0.81
Al_2O_3	13.9	15.7	15.9	16.8	4.0
Fe_2O_3	0.86	1.3	2.7	2.6	2.5
FeO	1.7	2.6	6.5	7.9	9.8
MnO	0.06	0.07	0.18	0.18	0.21
MgO	0.52	1.6	5.7	8.1	34.0
CaO	1.3	3.6	7.9	11.1	3.5
Na_2O	3.1	3.8	3.4	2.3	0.56
K_2O	5.5	3.1	1.3	0.56	0.25
P_2O_5	0.18	0.21	0.35	0.24	0.05

The wide variety of igneous rocks that make up the continental crust results from the simple fact that minerals do not all melt at the same temperature. Moreover, all components in an aggregate of different minerals begin to melt at a temperature that is lower than the melting point of any of the components on its own. You have learned from Block 3 that when peridotite from the mantle undergoes partial melting, magma of basaltic composition is formed. Clearly, partial melting is one way of providing a melt of different composition from that of its source. Equally, the composition of a magma can be changed during crystallization if early-formed crystals are removed from the melt. This process is known as fractional crystallization.

Both fractional crystallization and partial melting almost always result in melts that are more silica-rich than the parental magma and source rock respectively. This is because minerals rich in silica, such as quartz and feldspar, generally have lower melting points than less silica-rich minerals such as pyroxene and olivine. In Block 3, we have considered only three rock types: peridotite, basalt and its coarse-grained equivalent gabbro. These rock types make up the oceanic lithosphere. From Block 3, you should remember that gabbros and basalts are referred to as basic in composition, in contrast to peridotite which is ultrabasic since it has less than 45% SiO_2. Now we need to extend our classification to much higher silica contents. Igneous rocks with between 52 and 70% SiO_2 are termed **intermediate** in composition. From Table 4.1, you will see that andesites (diorites) and dacites (granodiorites) fall in this group. Rocks of the most silica-rich group have more than 70% SiO_2 and are referred to as **acid** in composition. Granite is the term for a coarse-grained rock of acid composition. The reasons for this nomenclature are given in Block 3, Appendix 2. Intermediate and acid igneous rocks form an important part of continental crust, and this Block examines how they form and what their significance is for the evolution of the continents.

Andesites (diorites) are found in island arcs and along active continental margins beneath which subduction occurs. Granites (rhyolites) are also found along active continental margins but unlike andesites or diorites are also found within continents, particularly where collision has occurred between two continental margins. Consequently, these tectonic environments hold the key to the formation of silica-rich magmas that characterize the continental crust.

The Block examines three tectonic settings in which acid and intermediate magmas are formed. Firstly, it examines island arcs, where one consequence of subducting oceanic lithosphere beneath oceanic lithosphere is the formation of andesitic (dioritic) magmas. Models for the origin of andesite are tested using techniques such as ternary phase diagrams and major-element and trace-element variation diagrams that have been introduced to you in Block 3.

Secondly, the Block examines the wide variety of igneous rocks formed at active continental margins such as the Andes. These include plutons of intermediate and acid compositions. The applications of Rb and Sr isotopes to dating the age of an igneous rock are described, together with their use for tracing the source from which magmas have been derived.

The third tectonic setting we examine is that of collision between two continents. We examine the consequences of thickening and heating the continental crust both in terms of metamorphism of minerals and in terms of forming new magmas.

This is an appropriate point to listen to the audio sequence 'Home Kit Rocks' (AV 07), which will provide you with a preview of the main rock types that will be discussed in the following Sections of the Block. The sequence lasts about 18 minutes.

4.2   ISLAND ARCS

The subduction of oceanic lithosphere can occur beneath either oceanic lithosphere or continental lithosphere. The former tectonic setting results in an island arc and the latter results in an active continental margin. (The term active continental margin distinguishes these environments from passive continental margins, such as along the east coast of the Americas, which are not underthrust by subduction zones.) This Section is concerned with magmatism at island arcs.

The majority of currently active island arcs are found in the western Pacific (Figure 4.1). Other examples include the Lesser Antilles in the Caribbean, and nearer home there is an active island arc in the Aeolian Islands of the Mediterranean (Plate 4.1 in the Colour Plate Section). Elsewhere on the planet, subduction of oceanic crust generally occurs beneath continental lithosphere, resulting in an active continental margin.

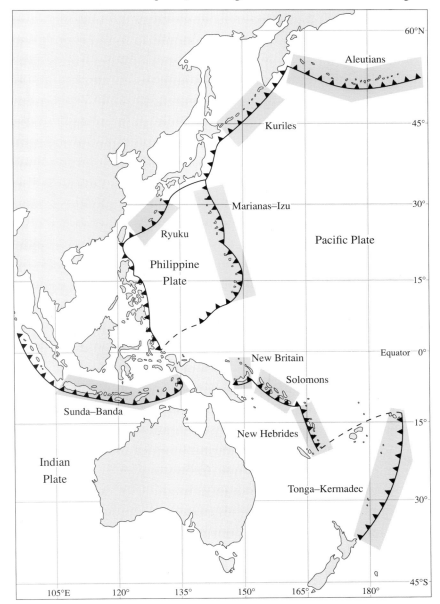

Figure 4.1   Distribution of currently active island arcs in the western Pacific.

The importance of island arcs lies in the intriguing possibility that they may represent sites of formation of new continental crust where the continents are being 'distilled' from the mantle. You may remember from Block 1 that the average composition of the upper continental crust is similar to that of a granodiorite. The average composition of the *entire* continental crust is similar to that of an andesite (diorite) and is consequently much more silica-rich than the basalts formed at ocean ridges. Somewhere on the planet, processes must be going on, or have gone on in past geological time, that transform mantle melts into the silicic rocks that make up the continents. Typical magmas found at island arcs are mid-way in composition between the basalts of the ocean floor and the granites of the continents. Is it at island arcs that we can see the process of making continents from the mantle, caught in its early stages?

4.2.1   ORIGIN OF MAGMAS AT ISLAND ARCS

Subduction at island arcs results in the subducted oceanic lithosphere being progressively heated both by friction along its upper surface, and by conduction as the slab descends deeper into the asthenosphere (Figure 4.2).

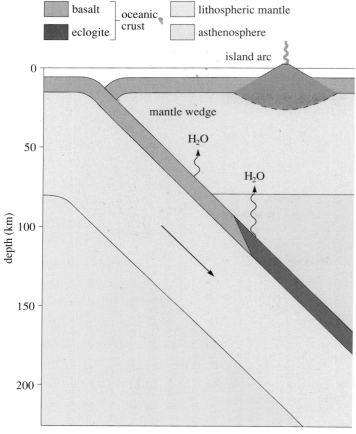

Figure 4.2 Vertical cross-section through a schematic island arc.

At depths of 50–100 km, the heating of the initially cold upper surface of the slab results in chemical reactions between minerals that release water, known as **dehydration reactions**. Such reactions occur both in the sediments in the upper layer of the subducted slab and also in the basalts of the underlying layer which contain hydrous minerals. The result of dehydration reactions is to drive off fluids into the prism of upper mantle above the subduction zone, known as the **mantle wedge** (Figure 4.2). The pressure on the slab also increases with depth, and at depths of about 100 km, both pressure and temperature are sufficiently high for mineral reactions to take place that replace the plagioclase feldspar and a sodium-rich pyroxene (which constitute the basalts and gabbros of oceanic crust) with garnet and pyroxene. The result is a dense metamorphic rock called eclogite, composed of red garnets and green pyroxenes (Sample 4 in the Kit).

The most conspicuous feature of island arcs is the magmatic activity that erupts at the surface, forming chains of volcanoes (Plate 4.1). In order to understand where this magmatism comes from, we return to the thermal structure beneath an island arc — a topic that you have met before in Blocks 1 and 2.

At its simplest level, the fact that magmas form at island arcs means that the solidus of the melting curve has been crossed at the source from which the magmas are derived. Melting in the mantle can be triggered by decompression as we know from Block 3, but beneath an island arc, the interaction between temperature and (water-rich) fluid is the key variable for generating melts. Figure 4.3 shows the variation of temperature with depth in the subducted slab and mantle wedge for a particular model of an island arc. Real island arcs may differ from this simple model in several important respects such as the angle or the rate of subduction or the thickness of the lithosphere.

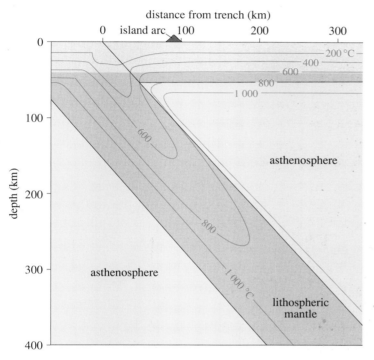

Figure 4.3 Thermal model for subduction of 100 km thick oceanic lithosphere at an island arc.

The temperature of melting for both basalt and peridotite under wet and dry conditions has been determined experimentally (Figure 4.4). You are familiar with the solidus of *dry* peridotite from Block 3, which is appropriate for magmas formed at spreading centres and hot spots.

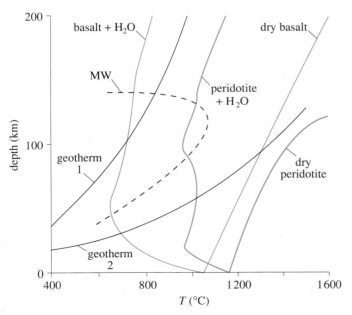

Figure 4.4 Solidus curves for melting of basalt and peridotite under wet (0.4% H_2O) and dry conditions. Two possible geotherms (1 and 2) are shown for temperatures in the upper part of the subducted slab. MW gives variation of temperature with depth in the mantle wedge.

In general, there is very little water available at depths within the Earth greater than a few tens of kilometres. However, as explained earlier, the subduction of an oceanic slab introduces water (as hydrous minerals) to great depths (up to 100 km). The release of this water through dehydration reactions allows wet melting to occur beneath island arcs.

There are two potential sites for melting at island arcs. Firstly, the upper surface of the subducted slab might become hot enough for partial melting to be possible. Secondly, the mantle wedge, above the subduction zone, might be heated sufficiently for partial melting to occur. Figure 4.4 shows two possible geotherms for the upper part of the subducted slab. Geotherm 1 models the temperature variation in a cold slab resulting from rapid subduction but ignoring frictional heating. Geotherm 2 is a warm slab model where friction is taken into account.

ITQ 1

Examine Figure 4.4 to determine whether the upper surface of the subducted lithosphere is likely to melt at depths of (a) 100 km, or (b) 200 km.

ITQ 2

Under what conditions could melting take place in the mantle wedge?

We can therefore conclude that melting of both the subducted slab, and the mantle wedge are possible at depths in the range of 100–200 km beneath island arcs because of the fluids* released from the dehydrating slab.

At greater depths, dehydration of the slab will be complete, and it will melt only if the frictional heating effects on the slab are extreme. This is an important conclusion because it tells us that the behaviour of fluids rather than increased temperatures or decreased pressures control melting beneath island arcs.

ITQ 3

Would you expect any difference between the compositions of magmas formed from melting the mantle wedge and those formed from melting the subducted slab?

The dominant rock-type found at island arcs is andesite or its plutonic equivalent diorite. The chemical compositions of basalts and andesites are different; for example, andesites have a higher silica content than basalts (Table 4.1). If you compare Samples 3 and 6 in the Kit, you will notice that andesites are lighter in colour than basalts because they contain a lower proportion of ferromagnesian minerals such as pyroxene and a higher proportion of light-coloured minerals such as feldspar. If we refer to Table 4.1 (or to Appendix 2, Block 3), we can see that andesites are intermediate in composition between basalts and granites. They differ in mineralogy from basalts in containing less pyroxene and no olivine. Instead, small quantities of quartz or alkali feldspar may be present, together with two minerals that you have not met before; brown mica (biotite) and amphibole (Appendix 1, Block 3). Both minerals contain

---

* In this Section, fluids means an H_2O-rich phase, *not* magma. In the final Section, fluids of other compositions (CO_2-rich) are also discussed.

water in their structure, and their presence in igneous rocks formed at island arcs is an indication of the water content of the magmas from which they crystallize.

Andesite also has a lower density than basalt, which is why old island arcs cannot be subducted. They form irreversible additions to the Earth's crust, and in this respect they are like true continental crust.

Although andesites are abundant at island arcs, they are not the only rock type to be formed in such an environment. Studies of several island arcs have revealed that the earliest magmatism was characterized by the eruption of basalt. There is no difficulty in accounting for basaltic magmas at island arcs. These can be generated by partial melting of the mantle wedge. But why are basalts found in young island arcs, whereas andesites are much more abundant in mature arcs?

The simplest explanation is that basalt is always formed above the subduction zone as the primary magma. This magma is modified during ascent by fractional crystallization. The crystallization and precipitation of a mineral with a low silica content, such as pyroxene or olivine, results in a melt with a higher silica content than the basaltic parental magma. In young island arcs, the basaltic magmas can rise to the surface unimpeded. However as the arc develops, a thick volcanic pile accumulates that makes it increasingly difficult for basaltic magmas to find escape channels directly to the surface. Instead they tend to accumulate in magma chambers near the base of the pile. Cooling within such chambers results in fractional crystallization to an andesitic composition.

The common association of basalt and andesite at island arcs does seem to suggest that the primary magmas at island arcs are basaltic and that andesites form by fractional crystallization of basalt. If we can prove this to be true by examining the appropriate phase diagrams and geochemical plots, then we can conclude that the source of island arc magmatism lies in the mantle wedge, not in the subducted plate.

4.2.2 THE PLAGIOCLASE-FELDSPAR SYSTEM

In order to evaluate the hypothesis that andesites result from fractional crystallization of basalt, we need to apply our understanding of the geochemical implications of fractional crystallization to the chemical variations observed between basalts and andesites at island arcs.

Basalt is made up of varying proportions of plagioclase and pyroxene, often with a small proportion of olivine. The Fo–An–Di ternary system we discussed in Block 3 (Section 3.6) is not appropriate for understanding the fractional crystallization of basalts because the binary system of plagioclase feldspar plays a key role in the evolution of more silicic magmas. Hence the results of fractional crystallization of a basaltic magma can best be understood by examining the ternary system made up of diopside (a clinopyroxene containing calcium and magnesium), albite (the sodium-rich plagioclase) and anorthite (the calcium-rich plagioclase). However, before diving into a ternary system that includes a solid-solution series, you need to become more familiar with crystallization in the binary system of plagioclase feldspar.

In Block 3 (Section 4.1.2), we considered partial melting in the olivine binary system. Plagioclase forms a more complex solid-solution relationship than olivine in that the plagioclase feldspars vary in composition between albite ($NaAlSi_3O_8$) and anorthite ($CaAl_2Si_2O_8$) by the coupled ionic substitution of (Ca^{2+},Al^{3+}) for (Na^+,Si^{4+}). The chemistry and structure of the plagioclase feldspars are summarized in Appendix 1 (Block 3).

11

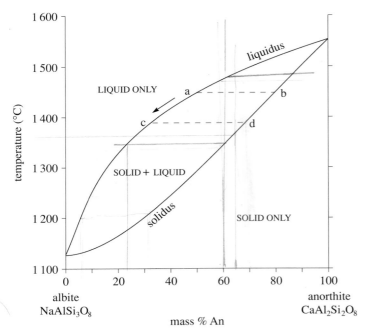

Figure 4.5 Phase diagram of temperature against composition for plagioclase feldspar determined experimentally at a pressure of $10^5 \, N \, m^{-2}$. The horizontal blue dashed lines are tie-lines linking the crystals and liquid that coexist on the solidus and liquidus respectively at a particular temperature.

The phase diagram for the plagioclase feldspar system is similar to that for the olivine system. In both cases, partial melting results in a liquid enriched in the low-temperature component, which in the case of plagioclase is albite. You can interpret crystallization in the plagioclase–feldspar system simply by applying what you have learned from partial melting in the olivine system (Section 3.4, Block 3). In order to confirm your understanding, it is most important that you attempt the following ITQ.

ITQ 4

A sample of molten plagioclase feldspar has a composition of An_{70}. Use Figure 4.5 to answer the following questions:

(a)   How much albite does this sample contain? (Give your answer in mass %.)

(b)   What is the temperature at which the first crystals appear?

(c)   What is the composition of the first crystals to form?

(d)   What is the composition of the total sample after half of it has crystallized?

(e)   At what temperature does the sample become completely solid?

(f)   What is the composition of the last drop of liquid before crystallization is complete?

(g)   What is the composition of the crystals in equilibrium with that last drop of liquid?

It may have puzzled you to note that the liquid and the crystals in the plagioclase system *both* manage to become richer in albite as the temperature falls. This is a result of the changing *proportion* of crystals and liquid. As the liquidus is crossed, only a small proportion of the sample is made up of crystals. By the time the solidus is crossed, there is virtually no liquid remaining.

Throughout this discussion, we have assumed that complete equilibrium was maintained between the crystals and the liquid. Let us look now at what happens if complete equilibrium does not occur.

4.2.3 EQUILIBRIUM OR FRACTIONAL CRYSTALLIZATION?

In our discussion of the plagioclase system (Figure 4.5), it was assumed that complete equilibrium was maintained between the crystals and the liquid. In other words, the compositions of the crystals, and of the liquid, would remain unchanged however long the temperature stayed constant. Chemical equilibrium is not a static situation, but a state of dynamic balance between opposing processes. Thus, if crystals and liquid are to remain in equilibrium in the plagioclase system, there must be continuous reaction between them, so that at any one temperature all the crystals have a unique composition given by the corresponding point on the solidus in Figure 4.5. For example, at $1445\,°C$, points a and b mark the composition of coexisting crystal and liquid compositions which are An_{80} and An_{50} respectively. This equilibrium can be expressed by the following equation:

$$An_{80}(solid) \rightleftharpoons An_{50}(liquid) \qquad \text{(Equation 4.1)}$$

However, this equilibrium is very sensitive to temperature, and thus, by the time the sample has cooled to $1385\,°C$, the equilibrium between solid and liquid is given by points c and d:

$$An_{68}(solid) \rightleftharpoons An_{32}(liquid) \qquad \text{(Equation 4.2)}$$

Thus the balance between the proportions of (Ca^{2+}, Al^{3+}) and (Na^+, Si^{4+}) in the crystals compared with that in the liquid has shifted dramatically with the fall in temperature from $1445\,°C$ to $1385\,°C$. Moreover, the fact that the liquidus and solidus are smooth curves on Figure 4.5 indicates that these changes have taken place *continuously* — we could write a different equilibrium equation for every temperature between $1445\,°C$ and $1385\,°C$.

For these exchange reactions to take place between the crystals and the liquid so that the composition of both can change continuously on cooling, the ions must be able to diffuse through the solid crystals.

Diffusion takes time, and if cooling is too fast, there may be insufficient time for the crystals to reach equilibrium with the liquid. All experimental systems such as Figure 4.5 are determined under equilibrium conditions for a fixed pressure. However, natural magmas may not maintain chemical equilibrium during cooling as we shall see.

The rate at which ions diffuse depends not only on the material through which they are passing, but also on the temperature and how the concentration of the element varies within the material. The relations between these factors are complex, but, for example, it would take about 40 000 years for the Ca in a crystal of plagioclase 1 cm in diameter to equilibrate with that in a well-mixed, homogeneous liquid at $1000\,°C$.

By contrast, if no liquid is present, adjacent minerals may not equilibrate with one another even after billions of years. This is particularly true at low temperatures, and it is just as well for those of us who wish to study igneous processes, since one consequence of extremely slow diffusion is that mineral compositions, which may reflect processes and reactions at over $1000\,°C$, are still preserved indefinitely for us to study at room temperatures.

If equilibrium has been maintained during cooling in the plagioclase system, then all the crystals should have the same composition and they should be homogeneous (that is, there should be no chemical variations between or within individual crystals).

Note that to describe a crystal as homogeneous does not necessarily imply that it is 'pure'. A crystal of 100% anorthite consists purely of anorthite and we may assume that it is homogeneous. However, a crystal of An_{50} is obviously made up of albite and anorthite (50% of each), and yet it is homogeneous if everywhere in the crystal the composition is An_{50}.

Homogeneous crystals are often found in large igneous intrusions which we know must have cooled slowly. However, in many igneous (particularly volcanic) rocks, plagioclase phenocrysts are often spectacularly zoned (Figure 4.6). Where such zoned crystals occur, it is fairly clear that the equilibrium between the liquid and the inner portions of the crystals has not been fully maintained during cooling. As the crystal grew, its surface was presumably in equilibrium with the liquid, but the inner parts of the crystal became isolated from the liquid and there was insufficient time for them to re-equilibrate with the liquid as the temperature fell. Equilibrium was only 'skin-deep'.

In summary, equilibrium is maintained by continuous reaction between crystals and liquid. Disequilibrium results if there is insufficient time for ions to move between the centre of crystals to the liquid. The shortage of time might reflect rapid cooling, or alternatively the crystals may have been removed physically before they had time to re-equilibrate with the liquid.

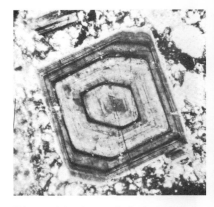

Figure 4.6   A zoned plagioclase phenocryst. The concentric lines represent boundaries between regions of different composition.

If the crystals are not in complete equilibrium with the liquid in a solid-solution series such as the plagioclase feldspars, then this affects the composition of both the crystals and the liquid. Under equilibrium conditions, complete crystallization results in homogeneous plagioclase crystals that have the same composition as the starting material. If we consider a liquid of composition An_{50} (Figure 4.7), the first crystals to form are of composition An_{80} (point b), and when the magma had cooled to 1 385 °C, the crystal composition has reached An_{68} (point d).

❑   What would happen to the composition of the liquid if crystals of composition An_{68} (point d on Figure 4.7) were then isolated and prevented from re-equilibrating with the liquid at lower temperatures?

■   The short answer is that, since these crystals are relatively rich in anorthite, their removal would cause the bulk composition of the sample to become depleted in anorthite (and hence richer in albite).

Hence, sluggish diffusion effectively isolates the cores of phenocrysts from the magma and thus changes the composition and behaviour of the magma at lower temperatures. This is a special case of a process we have met before, in Block 3 (Section 3.6.3) — fractional crystallization.

At 1 385 °C, about 50% of the sample An_{50} has crystallized under equilibrium conditions, the liquid has composition An_{32} (point c) and the crystals are An_{68} (Figure 4.7a). If equilibrium is maintained until crystallization is complete, then, as discussed earlier, the final sample will consist of homogeneous crystals of An_{50}. However, if at 1 385 °C, all the crystals (An_{68}) become isolated from the liquid and can no longer react with it, then they have no further role to play in the evolution of the liquid, whose composition is now An_{32}. *It is now as if we were considering the evolution of a bulk sample of An_{32}, rather than An_{50}.*

❑   Where in Figure 4.7 does a sample of An_{32} plot when the first crystals appear? What is the composition of those first crystals?

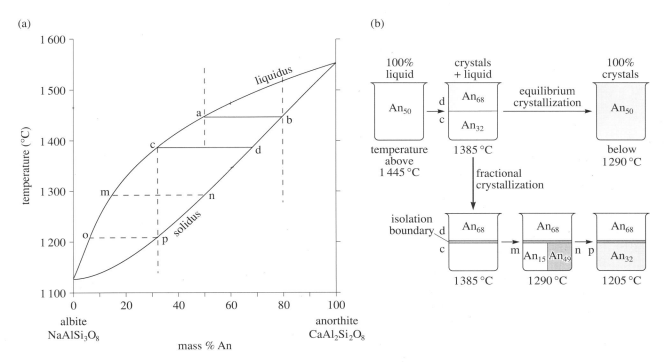

(a)

(b)

Figure 4.7   (a) Phase diagram of temperature against composition for the plagioclase feldspar system at a pressure of $10^5 \, \text{N m}^{-2}$. A sample of An_{50} crystallizes under equilibrium conditions until $1385\,°C$, whereupon the crystals (point d) become isolated from the liquid (point c) and cool separately. Blue horizontal lines are tie-lines between coexisting crystals and liquid, before (solid) and after (dashed) the crystals and liquid were separated at $1385\,°C$. (b) Diagrammatic sketch illustrating the crystallization of the sample of An_{50}.

■   The first crystals appear when the sample plots on the liquidus, that is, at c for sample An_{32}. The first crystals plot on the solidus at the same temperature, point d, An_{68} (Figure 4.7a). If equilibrium is then maintained, as sample An_{32} cools to $1290\,°C$, the liquid plots at m (An_{15}) and the crystals plot at n (An_{49}), and, when completely solid, the sample consists of homogeneous crystals of An_{32} (point p) at $1205\,°C$.

Reconsidering our initial sample of An_{50}, we can see that because crystals of An_{68} were isolated from the liquid when its composition reached An_{32}, we have ended up with a completely solid sample in which some of the crystals are of composition An_{68} and the rest are An_{32} (Figure 4.7b, second row).

> In summary, non-equilibrium crystallization in a system involving solid solution, such as the plagioclase feldspars, results in crystals with a greater range of compositions than that produced by equilibrium crystallization.

In this example, we have considered isolation of crystals from the liquid at a given temperature. While this can occur through physical removal, crystals can also be chemically isolated by rapid cooling which may result in zoned phenocrysts as seen in Figure 4.6.

Early crystals are richer in the higher-temperature end-member (anorthite) and are preserved because they do not re-equilibrate with subsequent (lower-temperature) liquids. Such liquids are therefore relatively impoverished in the high-temperature end-member (anorthite) and on solidification they result in crystals enriched in the low-temperature end-member (albite). By repeated crystallization and separation it is possible to arrive at very albite-rich liquids — and hence albite-rich crystals.

ITQ 5

Consider the crystallization of a sample of composition An_{60} using Figure 4.7.

(a)   Under equilibrium conditions, what are the compositions of the first and last crystals to form?

(b)   At 1 400 °C, the crystals and the liquid become separated physically and can no longer react with one another. Assuming that the remaining liquid then crystallizes under equilibrium conditions, what are the compositions of the different crystals when the total sample (An_{60}) is completely solid?

We have seen that, in general, fractional crystallization may take place for two reasons. Firstly, it occurs because the crystals are separated physically from the liquid. We have met this process in Section 3.6.3. Secondly, a rapid fall in temperature may result in the inner portions of individual phenocrysts not reaching chemical equilibrium with the later, lower-temperature liquids. This results in a range of compositions being preserved in a single crystal, as exemplified by the zoned plagioclase feldspar in Figure 4.6.

Now we have examined both equilibrium and fractional crystallization in the plagioclase feldspar system, we can extend our understanding to a ternary system that includes plagioclase.

4.2.4   THE DIOPSIDE–ALBITE–ANORTHITE TERNARY SYSTEM

The two essential minerals in basaltic rocks are Ca-rich pyroxene and Ca-rich plagioclase feldspar. If we confine our discussion to two-component systems, we are forced to compromise and to discuss basalts simply in terms of the phase relations of diopside and anorthite. We can now get closer to natural rocks by considering the ternary system diopside–albite–anorthite.

To revise your understanding of a simple binary system, you should attempt the following ITQ.

ITQ 6

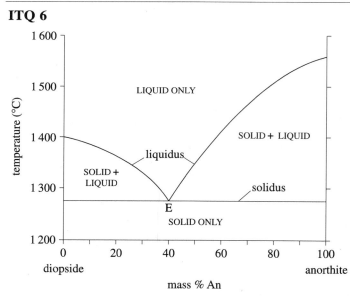

Figure 4.8   The diopside (Di)–anorthite (An) binary system

Using Figure 4.8, describe briefly what happens as a sample of 80% diopside and 20% anorthite is cooled from 1 600 °C to 1 200 °C. Pay particular attention to the temperature at which the first crystals appear and to their composition and also to the composition and temperature of the last drop of liquid before solidification is complete.

The diopside–albite–anorthite ternary system comprises two simple binary systems (diopside–anorthite and diopside–albite) and a solid-solution series (albite–anorthite).

Before proceeding further with this Section, it will be useful to review the video VB 04 'Melts to Minerals', which will introduce the ternary system diopside–albite–anorthite as well as revising what you have learned about ternary phase diagrams in Block 3. The video lasts 25 minutes.

In the relief models from your Kit, the vertical direction (the relief) represents temperature. Thus, as we discussed in Block 3 when considering the diopside–anorthite–forsterite ternary system, we can represent the liquidus surface by temperature contours on a plan view of the relief model (Figure 4.9).

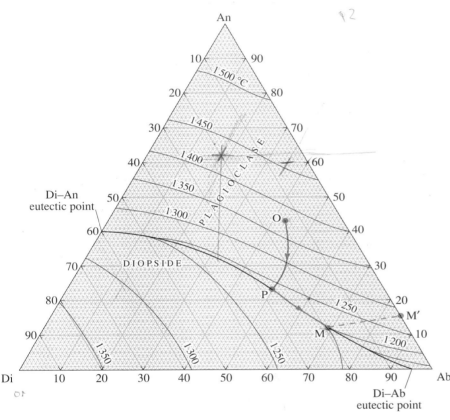

Figure 4.9 Diagram of the diopside–albite–anorthite system determined experimentally at pressures of $10^5 \, \text{N m}^{-2}$.

On these plan-view diagrams, in the same way as on topographic maps, the spacing between contours indicates how steeply the liquidus surface slopes. You should be able to assess the shape of the valley between the diopside and plagioclase fields by looking at the distribution of the temperature contours on Figure 4.9. If you find this difficult, compare Figure 4.9 with your three-dimensional block model of this system (Model 2); note in particular that the two ends of the valley are the eutectic *points* in the two-component systems of Di–Ab and Di–An, and that the valley floor slopes down (that is, down temperature) from the diopside–anorthite eutectic point to the diopside–albite eutectic point. The line of the valley floor is the **cotectic curve**, along which both diopside and plagioclase are in equilibrium with liquid.

This is an ideal point to listen to the second part of the audiovision sequence AV 08 'Phase Diagrams'. This part of the audiovision sequence lasts about 2 minutes.

17

We can now describe the crystallization path of a liquid as it cools and crystallizes in the diopside–albite–anorthite ternary system. For this, you will need to refer to Figure 4.9. We shall consider a liquid of 14% Di, 43% Ab and 43% An, and describe its crystallization path in a series of steps.

(i)   The liquid composition plots at point O on Figure 4.9. Since the sample is initially entirely liquid, it is at a higher temperature than the liquidus surface for the composition O (1 380 °C).

(ii)   As the temperature falls below 1 380 °C, crystals of plagioclase will form and the composition of the liquid will move away from the composition of the crystallizing plagioclase towards the cotectic curve because the liquid is becoming richer in diopside. The presence of diopside, while it will affect the temperatures of solidus and liquidus in the plagioclase binary system (Figure 4.5), does not affect the general behaviour of plagioclase crystallization. The compositions of plagioclase crystals become more albitic as crystallization proceeds and the liquid evolves along a curved path (O–P on Figure 4.9).

(iii)   At a temperature of 1 240 °C (P on Figure 4.9), the crystallizing plagioclase is joined by diopside (which has a fixed composition $CaMgSi_2O_6$). Plagioclase crystallizing with diopside will be richer in albite than plagioclase that crystallized at first.

(iv)   As the temperature falls below that at P, crystals of plagioclase and diopside form together. The composition of the diopside clearly remains fixed, but that of the plagioclase continues to become richer in albite. The composition of the liquid migrates along the cotectic curve towards M.

(v)   Under equilibrium conditions in the ternary system, crystallization stops when the plagioclase crystals have the same composition as those in the bulk sample (the initial liquid). The albite/anorthite ratio in the original sample was 1 (An_{50}), and plagioclase crystals of this composition are in equilibrium with a cotectic liquid composition at point M, corresponding to a temperature of about 1 200 °C. Thus, if chemical equilibrium has been maintained during cooling, the last drop of liquid has the composition M. Once that has crystallized, the sample has moved from the solid-and-liquid field to the solid-only field.

The solid end-product, after equilibrium crystallization, will be an aggregate of diopside and An_{50} plagioclase crystals. Moreover, the bulk composition of the solid end-product must of course be the same as that of O, the starting composition.

You may be wondering if the location of M on the cotectic curve can be determined graphically. To do this rigorously, we need to refer to experimental data for the plagioclase composition of interest. For example, laboratory experiments indicate that a mixture of plagioclase (An_{50}) and diopside, if heated, will start to melt at a temperature of 1 205 °C. The composition of plagioclase in this cotectic liquid is An_{16}. However, if we look at the plagioclase binary system (Figure 4.5), we see that plagioclase of composition An_{50} is in equilibrium with a liquid of composition An_{15}. We can conclude that adding diopside to the binary system makes only a small difference, in this case 1%, to the composition of the plagioclase component in the magma coexisting with plagioclase crystal of a given composition. The presence of diopside reduces both solidus and liquidus temperatures of the plagioclase binary system (Figure. 4.5) but does not significantly affect the shape of these curves. This conclusion allows us to obtain an approximate value for the composition of the final liquid during equilibrium crystallization by combining the ternary Di–Ab–An system with the binary plagioclase system, using the following procedure.

(1)   From the plagioclase binary system (Figure 4.5), read off the composition of the liquid composition that coexists with a solid of the starting plagioclase composition. For a solid of An_{50}, for example, the coexisting liquid has a composition of An_{15}.

(2)   Plot this composition on the Ab–An side of the ternary system (Figure 4.9). You need to measure the length of this face of the triangle, and for An_{15} plot the point 15% of this distance along the line from the Ab apex (M′ on Figure 4.9).

(3)   Determine the composition of the liquid on the cotectic curve with a plagioclase component of this composition. This may be obtained graphically by joining M′ to the Di apex. The point where this line crosses the cotectic curve (M) is the composition of the final liquid during equilibrium crystallization.

Hence, by combining information from Figures 4.5 and 4.9, we can estimate the final composition of the liquid during equilibrium crystallization for any starting composition in the system Di–Ab–An, although accurate determinations of liquid compositions must come from experimental data.

ITQ 7

A rock melt consists of 80% Di, 10% Ab, and 10% An. Use Figure 4.9 to predict the following (you may assume equilibrium crystallization throughout):

(a)   The temperatures at which crystallization will begin.

(b)   The sequence of mineral phases that form as the liquid cools.

(c)   The composition of the first and last plagioclase to crystallize from the liquid.

(d)   The composition of the final drop of liquid.

By using the Di–Ab–An system as an example, we have seen how the order in which minerals crystallize can be predicted depending on (i) the composition of the sample (the original melt), and (ii) the form of the experimentally determined phase diagram.

We appear to have deviated from the origin of andesites in order to explore the diopside–anorthite–albite ternary system, but in fact they are closely related. Equilibrium crystallization in this system drives the liquid to higher albite concentrations at the expense of diopside and anorthite. Now, since the concentration of SiO_2 in albite (68.6%) is higher than that in either diopside (55.5%) or anorthite (43.2%), this trend generates liquids of increasing silica contents. For example, compositions O, P and M in Figure 4.9 have silica contents of 56, 59 and 65% respectively.

If you refer to Appendix 2 (Block 3), you will see from their silica contents that O is a silica-poor andesite (properly called a basaltic andesite), P is an andesite and M a dacite, a composition between andesite and rhyolite. In this example, then, crystallization of a basaltic andesite results in a liquid of dacitic composition.

In general, basalts have compositions on the pyroxene side of the cotectic, whereas andesites have compositions along the cotectic curve between P and M (Figure 4.9).

But what happens to the composition of the liquid in the Di–Ab–An system if early-formed crystals are separated physically from the liquid and bulk equilibrium is not maintained? In other words, what are the consequences of fractional crystallization?

ITQ 8 gives you the opportunity to revise the effect of equilibrium and non-equilibrium crystallization in simple eutectic and solid-solution systems.

ITQ 8

(a)   In a binary eutectic system such as Di–An (Figure 4.8), is the composition of the liquid at any particular temperature different during non-equilibrium crystallization from its composition during equilibrium crystallization?

(b)   In the plagioclase system (Figure 4.7), will the lowest temperature liquid that can be formed by non-equilibrium crystallization (i) have the same composition, (ii) be richer in albite, or (iii) be richer in anorthite, compared with that resulting from equilibrium crystallization?

The answer to part (b) of ITQ 8 is most relevant to this discussion. Under equilibrium conditions, the last drop of liquid remaining from our original sample with 14% Di, 43% Ab and 43% An (point O) plotted at point M in Figure 4.9. M represents the composition of the liquid in equilibrium with plagioclase crystals similar in composition to that of the feldspar in the original sample, An_{50}.

However, during fractional crystallization, the early-formed plagioclase crystals no longer re-equilibrate with the later, lower-temperature liquids. Therefore

(i)   the remaining liquid becomes richer in albite (Figure 4.7);

(ii)   the liquid evolves further down the cotectic curve (on the low-temperature side of M, Figure 4.9).

We may conclude that *non-equilibrium crystallization in the ternary system Di–An–Ab drives the residual liquid further down the cotectic trough*, and under extreme conditions it may even reach close to the Di–Ab eutectic composition. Such a liquid contains well over 90% albite.

❑   Can you assess whether residual liquids become enriched or depleted in SiO_2 as the Di–Ab eutectic is approached?

■   Since albite contains more silica than anorthite, silica increases as the eutectic is approached, so the Di–Ab eutectic composition should have more SiO_2 than liquid M. In fact, the composition of the Di–Ab eutectic liquid has $SiO_2 = 68.2\%$.

Thus we arrive at a conclusion that is extremely important when we come to consider the chemical variations found in igneous rocks.

Fractional crystallization (that is, isolation of the early-formed crystals from the liquid) can result in a suite of liquids with a wide range of chemical compositions.

In the system Di–Ab–An, for example, fractional crystallization of basic magmas can yield magmas of intermediate compositions. Let us now see how these evariations compare with those observed in natural rocks.

4.2.5 THE LESSER ANTILLES: AN ISLAND-ARC CASE STUDY

The Lesser Antilles islands of the eastern Caribbean (Figure 4.10) are a well-studied example of an island arc: they are mainly volcanic and were formed by igneous processes above a subduction zone as we learned in Block 2 (Section 2.3.3). If you refer to the Smithsonian Map, you will find they are characterized by an arc of both shallow and deep-focus earthquakes, superimposed on an arc of volcanic activity. More detailed studies suggest two arcs with recent activity occurring along the younger arc (Figure 4.10). Volcanoes on some of the islands are still active (for example. Mt Soufrière on St Vincent erupted in 1979, and in 1902 the eruption of Mt Pelée on Martinique led to the loss of nearly 30000 lives). The need to understand volcanic processes, and particularly the risks to neighbouring population centres, has been at least in part responsible for the considerable amount of work that has been done in this area.

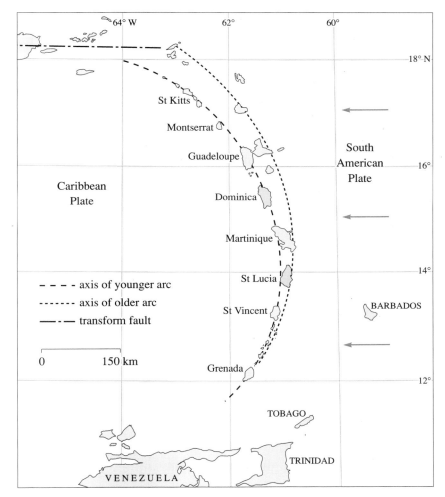

Figure 4.10 Sketch map of the Lesser Antilles island-arc in the eastern Caribbean. The blue arrows represent the relative movement of the South American plate presently being subducted under the Caribbean plate.

As our first example, we have chosen two volcanoes on the island of Dominica. Both of these volcanoes have been active in the last few million years, but while one erupted predominantly basaltic lavas the other appears to have produced andesites and rocks of higher SiO_2 contents. The two volcanoes are quite close to one another and by studying them it was hoped to ascertain whether the basalts and the andesites were related to one another and thus whether they could have been derived (by fractional crystallization) from the same parental magma.

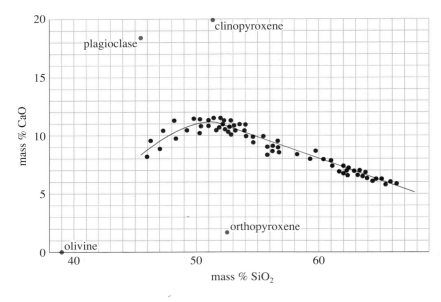

Figure 4.11 CaO versus SiO_2 concentrations from a suite of lavas and phenocryst minerals from two volcanoes on the island of Dominica. Blue lines mark smoothed compositional trends.

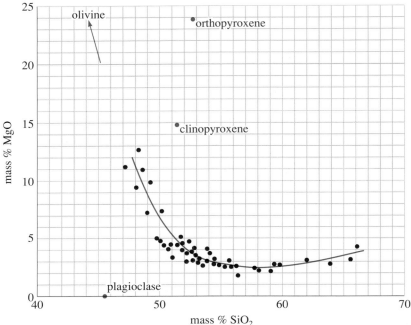

Figure 4.12 MgO versus SiO_2 concentrations from a suite of lavas and phenocryst minerals from two volcanoes on the island of Dominica. Blue lines mark smoothed compositional trends.

Figures 4.11 and 4.12 show SiO_2 plotted against CaO and MgO respectively, for a range of volcanic rocks from Dominica. Four minerals, plagioclase, clinopyroxene, orthopyroxene and olivine are commonly found as phenocrysts in the Dominica lavas and their average compositions are also plotted in Figures 4.11 and 4.12. Remember that *clinopyroxene* is the name given to those members of the pyroxene mineral group that are rich in CaO, whereas *orthopyroxenes* are poor in CaO (CaO < 2.5%). In both Figures 4.11 and 4.12, the trends are smooth and we can use the shape of such trends to infer what mineral or groups of minerals are fractionating from these magmas represented by the igneous rocks.

First, however, it is useful to compare briefly the compositions of the Dominica rocks with the calculated compositions of the liquids we have been discussing in the synthetic system Di–Ab–An. The liquids in the Di–Ab–An system varied from 56 to 68% SiO_2, which is similar to the higher SiO_2 rocks on Dominica, but the more silica-poor compositions cannot be reproduced in the ternary system. Moreover, although the system Di–Ab–An includes the two commonest calcium-bearing minerals found in basic igneous rocks (diopside and anorthite-rich plagioclase), it does not contain either of the Mg-rich minerals olivine or orthopyroxene. Since both olivine and orthopyroxene phenocrysts are observed in the Dominican lavas, we cannot use the Di–Ab–An ternary system directly to interpret their crystallization. Natural rocks are more complex than synthetic systems, and we turn first to the techniques we learned in Section 3.6.3 to interpret Figures 4.11 and 4.12.

❑ Can you observe any kinks in the trends defined by the Dominican data in Figures 4.11 and 4.12?

■ Yes, there appears to be a strong kink in the trends at about 50–52% SiO_2 on both diagrams, and a less obvious kink at about 60% SiO_2 on Figure 4.12.

❑ If the observed variations in SiO_2, CaO and MgO were controlled by fractional crystallization, what could you say about the composition of the crystals extracted (the *extract*) compared with that of the liquid? For samples with less than 50% SiO_2, would the extract contain more or less of the following constituents than the liquid: (a) SiO_2, (b) CaO, (c) MgO?

■ The trends of the data in the rocks with less than 50% SiO_2 (Figures 4.11 and 4.12) show an increase in CaO and SiO_2 and a decrease in MgO. The extract must therefore contain (a) less SiO_2, (b) less CaO, and (c) more MgO, than these liquids.

❑ What about the extract controlling liquids with 55–60% SiO_2?

■ The extract contains less SiO_2, but more CaO and just slightly more MgO than those liquids.

❑ Look at the trend of the rocks with less than 50% SiO_2 in Figure 4.11. Does the composition of any one phenocryst mineral plot on the trend in such a way that crystallization of that mineral will drive the composition of the liquid along that trend?

■ The removal of olivine would drive the composition of the liquids up a line of increasing CaO until it starts to flatten out at about 50% SiO_2. Similarly, olivine contains abundant MgO, and hence removal of olivine drastically reduces the MgO contents of the liquids with less than 50% SiO_2.

❑ What about the rocks with 52–60% SiO_2 — could their compositions reflect fractional crystallization of any one of the four minerals observed as phenocrysts?

■ The answer must be no. No one mineral plots on line with the trend of the analyses of rocks with more than 52% SiO_2. Thus, if this trend does reflect fractional crystallization, it must be due to the crystallization and removal of more than one mineral. We can see from Figure 4.11 that some combination of high-calcium minerals (plagioclase and/or clinopyroxene) and low-calcium minerals (olivine and/or orthopyroxene) is required.

In order to find out which of these minerals actually crystallized together in the Dominica volcanics, we must examine the rocks themselves. Phenocrysts represent minerals that were crystallizing before the liquid was cooled rapidly by extrusion on the Earth's surface. Thus, identification of the phenocryst minerals, and their proportions, provides us with an estimate of the bulk composition of the aggregate of minerals that were crystallizing together shortly before they were erupted. However, the proportions of minerals seen as phenocrysts may not be the same as the proportions extracted during a long period of fractional crystallization because some phenocrysts will have formed cumulates at the base and around the sides of the magma chamber, prior to eruption. Fortunately, we can use geochemical trends such as those of Figures 4.11 and 4.12 to test whether the suite of rocks from Dominica could be generated by fractional crystallization of the observed assemblage of phenocrysts.

A typical basalt from Dominica has three phenocryst minerals: olivine (10.5%), plagioclase (27.3%) and clinopyroxene (6.2%). These abundances are expressed as percentages of the *total* rock, but they may also be expressed as percentages of the phenocryst assemblage, which therefore consists of 24% olivine, 62% plagioclase and 14% clinopyroxene.

ITQ 9

The aggregate phenocryst assemblage (24% olivine, 62% plagioclase and 14% clinopyroxene) contains 14.2% CaO and 44.5% SiO_2. Plot that composition on Figure 4.11 and see where it plots in relation to the trend of the volcanic rocks with just 52–60% SiO_2. Could fractional crystallization of this phenocryst assemblage be responsible for the variations in CaO and SiO_2 in the lavas with 52–60% SiO_2?

The conclusion from ITQ 9 is that the trend of the chemical compositions of rocks with 52–60% SiO_2 can be produced by fractional crystallization of the minerals that are known to have been cystallizing from Dominica basalts when they erupted. Moreover, since the CaO and SiO_2 contents of the extract given in ITQ 9 lies on the trend of the lavas up to 66% SiO_2, we might argue that separation of these minerals was responsible for the composition of even the SiO_2-rich rocks. However, in so doing we would be falling into the trap of relying too heavily on the geometrical constraints of geochemical variation diagrams and would be ignoring what we know about the rocks themselves.

The phenocryst assemblage used in ITQ 9 was observed in a basalt and *it is therefore not going to occur in rocks of significantly higher SiO$_2$ contents*. High-SiO_2 rocks like rhyolites (the volcanic equivalent of a granite) largely consist of different minerals (quartz and feldspars) from basalts (olivine, pyroxene and plagioclase feldspar). Indeed, magnesium-rich olivine and quartz are never stable together in the same igneous rock. We also know from our own deliberations that the composition of plagioclase feldspar, for example, will change as the composition of the liquid changes: as more crystals form, both the liquid and the crystals become poorer in anorthite (see Figure 4.5). The plagioclase compositions plotted in Figures 4.11 and 4.12 are average values. Analyses of plagioclase feldspar from a rock of higher SiO_2 contents on Dominica gives 8.2% CaO, 58.6% SiO_2. If you plot this on Figure 4.11 you will find that it lies on the trend of the rock analyses, which suggests that separation of anorthite-poor plagioclase alone could control the compositions of the rocks with the highest SiO_2 content. The slight rise in MgO contents in the silica-rich samples (Figure 4.12) is consistent with this interpretation.

A second example is provided by olivine, which is abundant in low-silica rocks, but on Dominica is very rarely observed in rocks with more than 60% SiO_2, indicating that the liquids have moved out of the stability field of olivine. Instead, orthopyroxene starts to crystallize and, since like olivine it contains little CaO and significant amounts of MgO, the transition from olivine to orthopyroxene is not marked by any kink on the trends on Figures 4.11 and 4.12.

The gradual changes in the chemistry of the lavas at these two volcanoes on Dominica therefore suggest that they may well be related to one another. Moreover, by identifying and analysing the phenocryst minerals, we have deduced a fractional crystallization path capable of generating the chemical variations observed in the lavas. That is:

1   At less than 50% SiO_2, olivine crystallizes alone.

2   At 50–60% SiO_2, plagioclase and clinopyroxene crystallize together with olivine (at lower SiO_2 compositions) and orthopyroxene (at higher SiO_2 compositions). At this stage, the rocks formed would be similar to Sample 5, the basaltic andesite, in the Kit.

3   At more than 60% SiO_2, the amount of pyroxene crystallizing decreases until the chemical variation in very SiO_2-rich liquids appears to be controlled by crystallization of plagioclase alone.

Thus, we may conclude that the andesites on Dominica appear to have been derived by fractional crystallization from basaltic magma, which resulted from partial melting of upper mantle peridotite. We have seen how it was possible to argue from the observed geochemical trends that the role of plagioclase as a crystallizing phase became increasingly important in the more SiO_2-rich liquids. We have not been able to apply the Di–Ab–An ternary system directly to these rocks since the Dominica lavas are more complex than the synthetic system.

Before we leave the Lesser Antilles case study, we should examine a second suite of rocks made up largely of clinopyroxene and plagioclase that can be related to the Di–Ab–An ternary system we discussed in the previous Section. A volcano from the island of St Lucia provides rocks that vary from basaltic andesites (SiO_2 = 53%) to silicic andesites (SiO_2 = 61%). The more basic rocks have phenocrysts of just pyroxene, whereas more acidic (silica-rich) samples include both plagioclase and pyroxene phenocrysts. The compositions of the plagioclase phenocrysts show systematic variations through the series (Figure 4.13). Plagioclase phenocrysts in the less silica-rich andesites are zoned with a core composition of about An_{60} and a rim composition of about An_{45}.

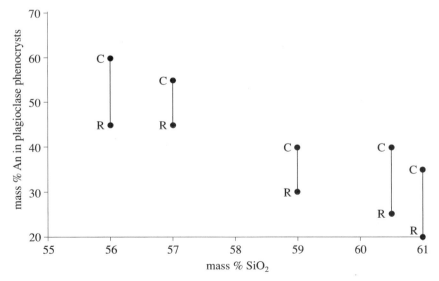

Figure 4.13   Variation of composition of plagioclase phenocrysts with silica content of bulk sample from a suite of lavas from St Lucia. C, core composition; R, rim composition.

In the more silica-rich rocks, plagioclase phenocrysts are also zoned, but there is a general trend towards albite-rich compositions for both cores and rims with increasing SiO_2. In the most silicic rocks, plagioclase cores have compositions of An_{35} with rims of An_{20}. Where zoning is less marked (i.e. core and rim compositions are similar), the plagioclase phenocrysts have reacted more extensively with the magma during cooling.

Since 95% of all minerals in this suite are clinopyroxene (which is close to diopside in composition) and plagioclase, we can interpret the crystallization history of these rocks with reference to the diopside–albite–anorthite ternary system (Figure 4.14). The early crystallization of clinopyroxene suggests the basaltic magma had an initial composition that lay within the diopside field (the actual composition of a basaltic andesite is plotted as point A in Figure 4.14). Progressive crystallization with cooling moved the composition away from the diopside apex towards point B on the diopside–plagioclase cotectic.

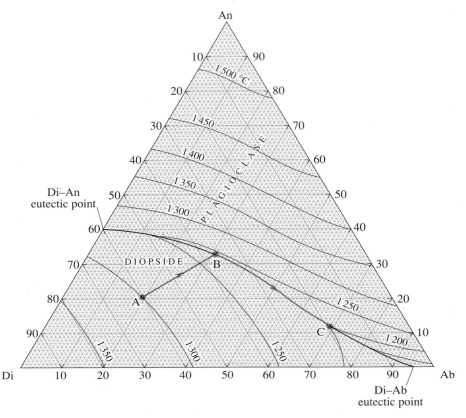

Figure 4.14   Phase diagram of the diopside–albite–anorthite ternary system with compositions A, B and C plotted from volcanic rocks from St Lucia.

Once plagioclase began to crystallize when the cotectic curve was reached at B, then the plagioclase phenocrysts either remain in suspension or would precipitate on the sides of the chamber as part of a cumulate. For crystals that remained in the magma, we know from Figure 4.5 that they would become more albitic as the temperature dropped, but that the slow rate of diffusion compared to crystal growth rates would lead to zoned crystals with anorthite-rich cores and albitic rims. On the other hand, removal of the crystals would drive the bulk composition of the magma down the cotectic towards the albite apex (Figure 4.14). Plagioclase from the most silica-rich and highly fractionated composition observed in the St Lucia suite (composition C, Figure 4.14) includes the most albitic plagioclase rims.

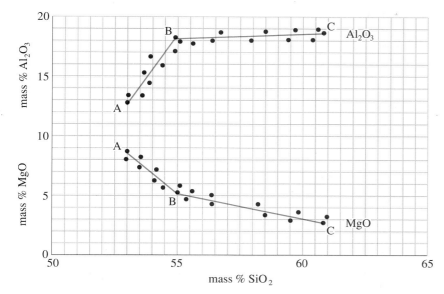

Figure 4.15   Al$_2$O$_3$ and MgO versus SiO$_2$ concentrations from a suite of lavas from St Lucia.

ITQ 10

Use the observed trends of Al$_2$O$_3$ and MgO versus SiO$_2$ (Figure 4.15) together with the compositions of diopside, albite and anorthite (Table 4.2) to confirm the petrological history of lavas from St Lucia that was deduced from the Di–Ab–An ternary system (Figure 4.14).

Table 4.2   Average composition (in mass %) of diopside phenocrysts and of pure albite and pure anorthite for use with ITQ 10.

	Di	**Ab**	**An**
SiO$_2$	51.1	68.6	43.2
Al$_2$O$_3$	2.0	19.4	36.6
MgO	14.9	—	—

From your answer to ITQ 10, you should conclude that the fractional crystallization path of natural rocks can be described with reference to the Di–Ab–An system provided the rocks are made up largely of a diopside-rich clinopyroxene and plagioclase. Most natural igneous rocks have a more complex mineralogy which often requires not only major element data, but also trace-element data, to unravel.

There is, however, an important constraint that restricts the application of experimentally determined binary and ternary systems to natural rocks. Phase relations are sensitive to pressure as well as temperature, and all the systems we have described so far have been determined at a fixed pressure (usually 10^8 N m^{-2} for the systems described in this Block). Before applying these diagrams to natural rocks, we have to be sure the difference between the pressure of fractional crystallization in the magma chamber and the pressure at which the phase relations in the appropriate diagram were obtained will not affect our interpretation. We can use this constraint to our advantage. For example, in both examples of fractional crystallization described from the Lesser Antilles, plagioclase crystallization has been important. Plagioclase is only stable as a crystallizing phase at depths of less than about 30 km. Since the crust beneath the Antilles arc is at least 30 km thick, we can conclude that fractional crystallization of the basalts to form andesites took place in the crust and not in the mantle. We can therefore infer that beneath island arcs, magma chambers develop within the crust where fractional crystallization of basaltic magmas leads to lavas of intermediate compositions.

4.2.6   TRACE ELEMENTS IN ISLAND-ARC MAGMATISM

So far, we have examined the relationship between mineralogy and major element chemistry in island-arc magmas. Now we turn to the information that *trace elements* give us on the processes that occur at island arcs.

The relationship between the amount of partial melting of a rock (F), the bulk partition coefficient (D) of a trace element between the residue and the liquid, the concentration of the trace element in the liquid (C_1) and in the source rock before melting (C_0) is given by

$$C_1 = \frac{C_0}{D + (1 - D)F} \qquad \text{(Equation 4.3)}$$

as described in Section 3.7.2, Block 3. The effects of fractional crystallization on the trace-element concentrations may also be quantified. If equilibrium crystallization occurs, then the same equation will describe the distribution of trace elements between crystals and melt. If you have trouble accepting this, then imagine a rock composed entirely of feldspar which is heated until it is 50% melt ($F = 0.5$). The trace-element concentrations in melt (C_1) and in the source (C_0) will be described by Equation 4.3. Now heat the rock until it is entirely molten, and proceed by cooling until 50% has crystallized out. Clearly, provided the crystals and liquid are in equilibrium, exactly the same situation arises as for partial melting and Equation 4.3 will define C_1/C_0 as it did for partial melting.

However, if non-equilibrium crystallization occurs by removal of the crystals from the melt, then Equation 4.3 can no longer be applied. Instead, C_1 and C_0 are related by

$$C_1 = C_0 F^{(D-1)} \qquad \text{(Equation 4.4)}$$

where C_1 is the concentration of the trace element remaining in the liquid and F is the mass fraction of melt remaining. As before, C_0 is the initial concentration of the trace element in the magma and D is the bulk partition coefficient of the trace element between crystals and liquid. You will not need to manipulate Equation 4.4 in this Course but you should notice that Equations 4.3 and 4.4 are different in form.

If only one mineral is crystallizing, then the bulk partition coefficient (D) equals the partition coefficient (K_D). The following three conclusions summarize *qualitatively* the implications of fractional crystallization for the trace-element concentrations of the melt.

(1)   For K_D less than 1, the crystallization of such minerals *increases* the concentration of the trace element in the liquid. Trace elements with small partition coefficients tend to remain in the liquid, and so, as crystallization proceeds, their concentrations will increase simply because the amount of liquid is being reduced.

(2)   If K_D equals 1, the concentration of the trace element in the mineral is the same as in the liquid and so fractional crystallization of that mineral will not affect concentrations of the trace element in the liquid.

(3)   When K_D is greater than 1, the concentration of that trace element in the liquid is *reduced* by crystallization.

From Section 4.1, we know that the predominant magma composition found at island arcs is andesitic. From considerations of the major elements, we know that the andesites of the Lesser Antilles can be generated from the coexisting basalts by fractionating the minerals olivine, plagioclase and clinopyroxene (ITQ 7). Do the trace-element variations in the rocks support such a model? Table 4.3 shows the mineral-liquid partition coefficients for several trace elements for six important rock-forming minerals. These values differ from those provided in Block 3 as the K_D values in Table 4.3 are determined for magmas of intermediate, rather than basic, compositions.

Table 4.3   Selected trace-element partition coefficients K_D = concentration of mineral in liquid of andesitic composition.

	Rb	**Sr**	**Ba**	**Zr**	**Y**
olivine	0.001	0.001	0.001	0.01	0.01
orthopyroxene	0.003	0.05	0.13	0.08	0.45
clinopyroxene	0.001	0.07	0.001	0.25	1.0
garnet	0.01	0.02	0.02	0.5	11.0
plagioclase	0.04	4.4	0.3	0.03	0.06
amphibole	0.01	0.02	0.04	1.4	2.5

ITQ 11

Would you expect Zr and Y to increase or decrease during fractional crystallization of each of the minerals olivine, plagioclase and clinopyroxene?

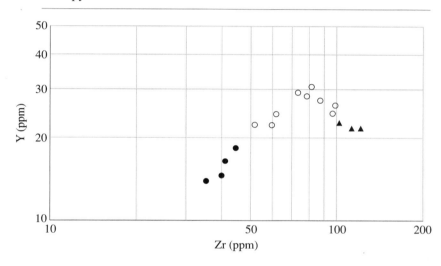

Figure 4.16   Plot of Zr against Y for lavas from Dominica: ●, basalts; ○, andesites; ▲, dacites. Note scales of both axes are logarithmic.

Figure 4.16 shows the observed trend of Zr–Y variation in basalts, andesites and dacites of the two Dominican volcanoes discussed in the previous Section. It is clear that fractional crystallization from the parental basaltic magma results in the enrichment of Zr and Y. However, within the field of andesitic compositions, there is a kink in the fractional crystallization path. What could cause this?

It is apparent that a new mineral was crystallizing that removed Y (but not Zr) from the melt. In other words, the bulk partition coefficient for Y was greater than one. An additional mineral found in Dominican andesites and dacites (but not in the basalts) is amphibole. Amphibole is a hydrous mineral, more common in granites than in andesites, and only rarely found in basalts (Appendix 2, Block 3). It was not discussed in Section 4.2.5 because fractional crystallization models based on major elements did not require additional minerals to explain the observed trend. This is partly because clinopyroxene and amphibole have similar major element compositions, and so fractional crystallization of the two minerals cannot be easily distinguished on simple major element plots. But Figure 4.16 shows us that the trace-element data do require a new mineral to be considered that has a high partition coefficient for Y. Only amphibole and garnet satisfy this condition (Table 4.3), and garnet is unlikely to play a role since it is not present in these volcanic rocks. We must conclude that amphibole fractional crystallization did occur within the andesites and dacites, but not in the early stages of fractional crystallization from the parental basalt.

Trace elements not only help us refine our understanding of the fractional crystallization history of a suite of igneous rocks, but can also provide critical information on the nature of the source region of parental melts. In Block 3, we learned how certain trace elements, such as Zr and Y, which are incompatible with mantle minerals, can discriminate between MORB and OIB magmatism. Such plots are also much used by geologists who are working on basalts where the regional tectonic setting is poorly understood. In such cases, the trace-element ratios, together with other geological evidence, can identify the likely tectonic regime at the time the basalts were erupted. We can now extend this approach to include a third category, **island-arc basalts (IAB)**.

MORB and IAB are characterized by critical differences in their trace-element concentrations. Figure 4.17 shows the ratio between a typical IAB and an average MORB composition for selected trace elements. It is clear that the elements Rb, Ba and Sr are strongly enriched in IAB, whereas the elements Zr and Y are slightly depleted, relative to the compositions of MORB. In both types of basalts, we are considering only primitive magmas and so these differences cannot simply be the result of fractional crystallization.

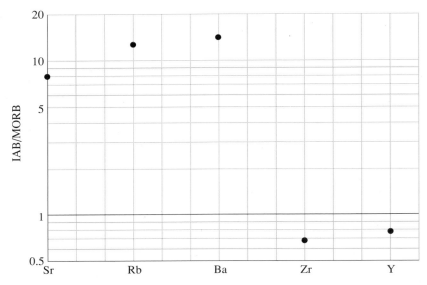

Figure 4.17 Trace-element concentrations of a typical IAB divided by concentrations of an average value for a MORB.

Rb^+, Ba^{2+} and Sr^{2+} are ions of low valency and large diameters that are highly soluble in aqueous solution. In contrast, Y^{3+} and Zr^{4+} form smaller ions with higher valencies that are insoluble in aqueous solution and hence are relatively immobile in the presence of a hydrous fluid. Basaltic magmas at island arcs must form in the mantle wedge since this is the only site where peridotite, the source material for basalt, can be melted (ITQ 2). When ocean floor is subducted, it is heated, and fluids are consequently released into the mantle wedge during dehydration reactions of minerals such as micas and amphiboles that contain water (Section 4.2.1). The enrichment of IAB in elements such as Rb, Sr and Ba is interpreted as the result of aqueous fluids being released into the mantle wedge during subduction, and thus enriching the source region of IAB in such elements.

The trace-element discriminant diagram is already familiar to you from Block 3. We can now extend it to include IAB (Figure 4.18). Although there is considerable overlap between IAB and MORB compositions, many IAB differ from MORB compositions in having lower Zr concentrations, and, in some cases, lower Zr/Y ratios.

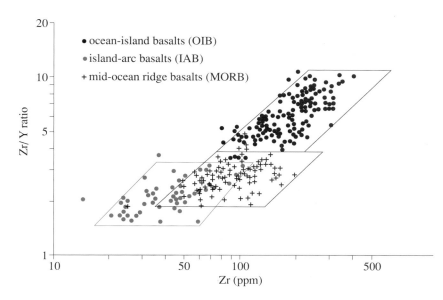

Figure 4.18 Zr/Y ratio plotted against Zr for basalts from three tectonic environments: MORB, OIB and IAB.

We know from Block 3 (Section 3.7.1, Table 3.7) that for common minerals in the mantle (olivine and pyroxene) the partition coefficient for Zr is small ($\ll 1$). If we refer back to the equation that defines the behaviour of trace elements during partial melting (Equation 4.3), we can see that

$$C_1 = \frac{C_o}{D + F(1-D)}.$$

For very low bulk partition coefficients ($D \approx 0$), then

$$C_1 \approx \frac{C_o}{F} \quad \text{or} \quad F \approx \frac{C_o}{C_1}.$$

This tells us that the melt fraction (F) is approximately inversely proportional to the concentration of highly incompatible elements in the melt. So we may infer from Figure 4.18 that the low Zr contents of some IAB reflects *higher* melt fractions in the mantle wedge than beneath ocean ridges. This is primarily a consequence of the fluids derived from the subducted plate migrating into the mantle wedge and triggering a higher melt fraction (by reducing the temperature of the solidus) than would be possible at an oceanic ridge. This simple explanation for the empirical relationship of incompatible elements in basalts to their tectonic setting (Figure 4.18) highlights yet again the critical role played by fluids in generating magmas at island arcs.

SUMMARY OF SECTION 4.2

- The compositions of lavas erupted at island arcs are predominantly, but not exclusively andesitic. Both basaltic and dacitic magmas are also erupted.

- Thermal models of island arcs suggest that melting may occur either in the upper part of the subducted lithosphere or in the overlying mantle wedge by partial melting of peridotite in the presence of H_2O that is expelled from the subducted plate through dehydration reactions at depths of 100–200 km.

- Fractionation of basalt can be modelled by the ternary diopside–albite–anorthite phase diagram. The principles controlling the evolution of both crystals and liquid are the same as those for the simpler two-component systems.

- If chemical equilibrium is not maintained between the crystals and the liquid, the later liquids are *depleted* in the high-temperature end-member and correspondingly *enriched* in the low-temperature end-member relative to liquids formed by equilibrium crystallization.

- Fractional crystallization may take place for two reasons: either the temperature falls too rapidly for the inner portions of individual crystals to re-equilibrate with the later, low-temperature liquids (resulting in zoned crystals), and/or the crystals and the liquid are simply separated physically as cumulates.

- Studies of major-element trends and phenocryst assemblages of volcanic rocks from the Lesser Antilles island arc suggest that the andesites and the higher-SiO_2 rocks were derived by fractional crystallization of a basaltic liquid. Since basalt is in its turn derived by partial melting of peridotite, we conclude that the andesites were not obtained by melting subducted ocean-floor basalt, but appear to have been derived from partial melts of peridotite in the mantle wedge.

- Abundances of trace elements with large ionic size and low valency (such as Rb, Sr and Ba) in IAB are enriched, compared to their abundances in MORB. Such elements are mobilized by aqueous fluids and their enrichment in IAB provides chemical evidence for the influx of fluids, derived from the subducted lithosphere, into the mantle wedge beneath island arcs.

- The low Zr concentration observed in some IAB suggests that a higher melt fraction is generated in the mantle wedge of island arcs than from the mantle beneath ocean ridges.

OBJECTIVES FOR SECTION 4.2

When you have completed this Section, you should be able to:

4.1   Recognize and use definitions and applications of each of the terms printed in the text in bold.

4.2   Identify possible source regions for magma generation at an island arc and recognize the distinct magma composition that can be formed at such source regions.

4.3   Describe both equilibrium and fractional crystallization of a liquid composition in a binary solid-solution system and in a ternary system in which solid solution occurs between two end-members.

4.4   Describe, using chemical variation diagrams, the relationship between the compositions of the crystallizing minerals and that of the evolving liquid.

4.5   Assess from variations in both major and trace-element compositions whether individual rocks from an igneous suite could be related to one another by fractional crystallization.

4.6   Distinguish the tectonic settings in which basalts are formed on the basis of the trace-element discriminant diagram.

Apart from Objective 4.1, to which they all relate, the eleven ITQs in this Section test the Objectives as follows: ITQs 1–3, Objective 4.2, ITQs 4–8, Objective 4.3, ITQs 9–11, Objectives 4.4 and 4.5.

You should now do the following SAQs, which test other aspects of the Objectives.

SAQS FOR SECTION 4.2

SAQ 1 (*Objective 4.3*)

The following questions are about the effects of fractional crystallization in the Di–Ab–An system (Figure 4.9).

(a)   Does fractional crystallization of diopside and plagioclase drive the liquid along the cotectic curve towards the Di–Ab or the Di–An eutectic?

(b)   Does fractional crystallization in the Di–Ab–An system result in liquids richer or poorer in CaO than those generated by equilibrium crystallization?

(c)   Given the SiO_2 contents of diopside, albite and anorthite listed in Table 4.2, is the composition of the liquid at P (Figure 4.9) richer or poorer in SiO_2 than that at the Di–Ab eutectic?

Table 4.4 SiO_2 and Al_2O_3 contents (in mass %) of selected volcanic rocks (for use with SAQ 2).

Analysis	SiO_2	Al_2O_3
14	46.8	16.0
24	49.4	16.5
31	56.1	17.2
34	58.3	18.5
27	60.6	17.4
17	62.3	17.3
39	63.4	16.2
19	70.3	13.0
25	73.0	11.5

SAQ 2 (*Objective 4.4*)

Table 4.4 presents the SiO_2 and Al_2O_3 contents of a suite of volcanic rocks. Plot these data on a graph of Al_2O_3 against SiO_2 (Figure 4.19), join up the points by a smooth curve and then answer questions (a), (b) and (c).

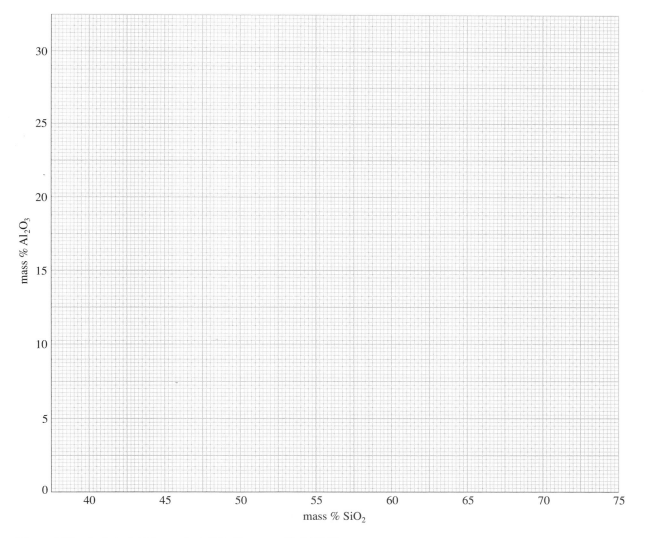

Figure 4.19   A plot of Al_2O_3 against SiO_2 (for use with SAQ 2).

(a) Describe briefly the shape of the curve you have plotted on the graph of Al_2O_3 against SiO_2. Is it straight or kinked?

(b) In terms of fractional crystallization, what is the significance of a change in slope in the trend of the analytical data plotted on chemical variation diagrams such as Al_2O_3 against SiO_2?

(c) Assume that these rocks are related by fractional crystallization. Would you expect the Al_2O_3 and SiO_2 contents of the separated minerals (the extract) to be higher or lower than the liquids at (i) 47% SiO_2, and (ii) 60% SiO_2?

SAQ 3 (*Objectives 4.4 and 4.5*)

Plot the SiO_2 and Al_2O_3 contents of the minerals in Table 4.5 on the same graph as the results from SAQ 2.

Could the evolution of the liquid from 47 to 56% SiO_2 be due to the crystallization and separation of any *one* of the minerals in Table 4.5 by itself?

Table 4.5 SiO_2 and Al_2O_3 contents (in mass %) of selected minerals from the volcanic rocks of Table 4.4 (for use with SAQ 3).

Mineral	SiO_2	Al_2O_3
Olivine	39.0	0
Plagioclase	50.0	31.8
Pyroxene	52.0	1.5
Alkali feldspar	63.5	20.6

SAQ 4 (*Objective 4.4 and 4.5*)

In the volcanic rocks from Table 4.4 with more than 60% SiO_2, Al_2O_3 starts to decrease as SiO_2 increases.

(a) How does this differ from the trend in the rocks with less than 58% SiO_2?

(b) Can you suggest an explanation for this change in slope (or kink) in the trend of the analytical data at about 58% SiO_2?

(c) Which mineral, or minerals, appears to influence most the evolution of the liquids from 60 to 73% SiO_2?

SAQ 5 (*Objective 4.6*)

Two basalts have the following trace-element concentrations: basalt A, Zr = 30 ppm, Y = 15 ppm; basalt B, Zr = 200 ppm, Y = 67 ppm.

(a) What can you infer about the tectonic setting in which they formed.

(b) Which do you think would have a higher Rb concentration? (Provide a one-sentence explanation for your deduction.)

4.3 ACTIVE CONTINENTAL MARGINS

The difference between island arcs and active continental margins is simply that subduction occurs beneath oceanic lithosphere at island arcs, but beneath continental lithosphere at active continental margins (Figure 4.20). By far the best documented example of a contemporary active continental margin runs along the western coast of South America (Block 2, Section 2.5.3), and for this reason active continental margins are sometimes also known as Andean margins after the South American mountain range.

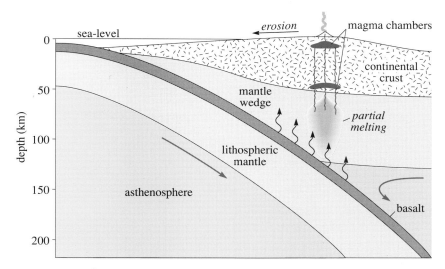

Figure 4.20 Vertical cross-section through a schematic active continental margin. (Arrows in the asthenosphere represent convection movements.)

The style of magmatism is quite distinct at active continental margins compared with island arcs. This can be illustrated by a sketch section across the igneous rocks of Peru that have been emplaced along the west coast of South America, immediately above the subducted Nazca plate (Figure 4.21). The abundance of plutonic, rather than volcanic, rocks in the section is partly a result of the deeper levels of erosion exposed in the Peruvian batholith which in turn is a consequence of rapid uplift of thick continental crust. Active and recently active volcanoes are common elsewhere along the Andean chain, particularly further south in Chile (see Plate 4.2 and the Smithsonian Map).

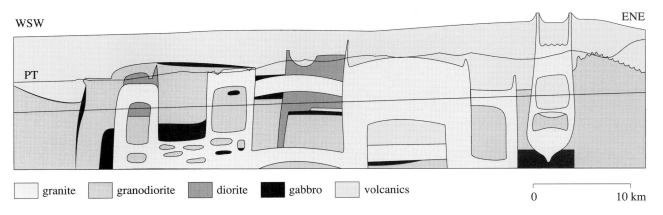

granite   granodiorite   diorite   gabbro   volcanics

0   10 km

Figure 4.21 Cross-section of the igneous rocks of Peru before erosion. PT indicates present topography. The Pacific coastline lies to the left of the diagram. (Vertical exaggeration ×2.)

One feature apparent from Figure 4.21 is that magmas of basaltic composition (gabbros) are less abundant than plutons of intermediate (diorite or granodiorite) or acid (granite) compositions. With increased silica contents, plutonic rocks contain progressively more quartz and alkali feldspar, and a lower proportion of ferromagnesian minerals such as pyroxene and amphibole. In granites, biotite is usually the commonest dark mineral (Sample 7 from your Kit) which may be present with either amphibole or a white mica called muscovite. Muscovite is structurally similar to biotite, but differs chemically in that Fe^{3+}, Fe^{2+} and Mg^{2+} ions are replaced by Al^{3+} (Appendix 1). The relationship between mineralogy and silica content for all these rock-types is given in Appendix 2 (Block 3). Volcanic equivalents of granites (rhyolites) occur as pumice and lava. Because of the high viscosity of silica-rich magma, the nucleation and growth rates of crystals in a rhyolite liquid can be slower than the rate of cooling after eruption, resulting in a glassy texture. Such a rock is known as obsidian, and an example is provided in the Kit (Sample 8).

Chemically, the plutonic rock series from diorite through granodiorite to granite shows a steady increase, not only in SiO_2 but also in alkali elements such as sodium and potassium (Table 4.1). Trace elements like Th and U also increase with SiO_2. Since the three elements (K, Th and U) that provide the Earth's important heat-producing radioactive isotopes are all enriched in silica-rich magmas, the chemistry of the continental crust provides a simple explanation for the fact that geothermal gradients are much steeper within continental crust than within oceanic crust (Block 1, Section 1.12).

In many respects, the potential source regions for magmas at active continental margins are similar to those we have discussed at island arcs. Subduction of the oceanic lithosphere will result in fluids being released into the mantle wedge which will allow partial melting (Figure 4.20). However, there is an added variable to the geochemical evolution of magmas, for now there is continental crust that may also contribute to magmatism. Firstly, it might be heated and partially melted by the rising basaltic and andesitic magmas. This is likely to occur in magma chambers within the continental crust where the rising magmas might be trapped, and subsequently cool. At this stage, both fractional crystallization and melting of the wall-rocks of the magma chamber might occur. Secondly, the continent will be eroded and continent-derived sediment deposited off-shore. This detritus could be subducted with the oceanic crust and enter the source region from below. Before we can evaluate how much magmatism at active continental margins comes from remelting old continents, and how much comes from the mantle below, we must address the problem of the origin of granite.

4.3.1 ORIGIN OF GRANITE

Ultimately, all rocks are traced back to the mantle, although in some cases many cycles of remelting, crystallization, erosion and deposition may be involved. However, can granites ($SiO_2 > 70\%$) be formed directly from the mantle ($SiO_2 < 45\%$) by fractional crystallization of a mantle-derived melt?

We have already looked at the effects of fractional crystallization in the Di–Ab–An system and generated dacitic liquids ($SiO_2 = 68\%$) from a basaltic andesite ($SiO_2 = 56\%$) in Figure 4.9.

ITQ 12

Given that the SiO_2 content of the Di–Ab eutectic composition is 68%, does this mean it is impossible to produce a granite liquid of $SiO_2 = 74\%$ by fractional crystallization of a basalt?

Before we examine an appropriate ternary system that will demonstrate how high-silica magmas can form, we must ask a more general question. Continental crust has an average overall composition close to that of an andesite ($SiO_2 = 60\%$). How can large volumes of magma of this composition be formed at an active continental margin?

If we assume that the starting liquid composition (basalt) had a silica content of 50% and the average silica composition of the cumulates removed from the melt is 44.5% SiO_2 (as in ITQ 9), we can easily calculate the proportion of final liquid of andesitic composition to initial basaltic liquid (F).

The silica content of the basalt must equal the silica content of the final liquid plus that of the cumulates removed from the melt (i.e. the mass balance is preserved during fractional crystallization).

Silica content of 100 kg of basalt = 50 kg.

Silica content of remaining liquid after fractional crystallization = 60F kg.

Silica content of cumulates = 44.5$(1 - F)$ kg.

Hence, $50 = 60F + 44.5(1 - F)$, or $F = 0.35$.

This means that the mass proportion of the initial basaltic liquid that must be extracted as cumulates during fractional crystallization is 1 − 0.35 or 0.65. In general, we can say that about 60–70% of a basaltic liquid must be extracted as cumulates to leave a liquid of andesitic composition. This means there are two noteworthy consequences of the hypothesis that average continental crust has been formed by fractional crystallization of basaltic liquids derived by partial melting of upper mantle peridotite above subduction zones:

1   The crystals that become separated from magmas during fractional crystallization presumably collect somewhere. If 70% fractional crystallization takes place, then there should be about 2.3 km³ of crystals for every cubic kilometre of andesite liquid (about 60% SiO_2) if we ignore any differences in density between crystals and magma.

2   Basaltic liquid probably represents about a 20% melt of peridotite: therefore for each cubic kilometre of andesite liquid, there should also be about 13.3 km³ of residual solid material left behind in the upper mantle.

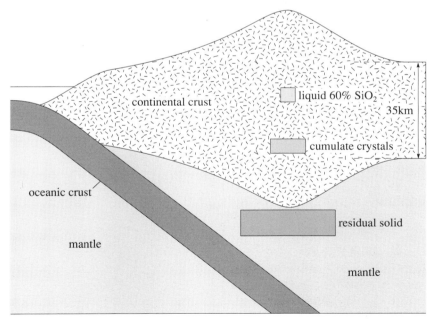

Figure 4.22 A schematic section through an active continental margin such as in the South American Andes. For discussion see text.

This model is illustrated in a schematic sketch section through a continental destructive plate margin (Figure 4.22). Continental crust is normally 30–35 km thick, but it may increase up to 70 km beneath areas of high mountains like the Andes and the Himalayas. For the Himalayas, this unusual thickness is attributed to collision between two continents as we know from Block 2, whereas in the Andes, it is thought to reflect the addition of new material (magma) to the crust above a subduction zone. In Figure 4.22, the size of the 'boxes' illustrates the relative quantities of andesite liquid, cumulate crystals and residual solid material left behind in the upper mantle according to the model that we outlined above. Thus, $333 \, km^3$ of residual upper mantle material and $58 \, km^3$ of crystals result from the generation of just $25 \, km^3$ of intermediate liquid — equivalent to the size of a small andesitic volcano.

The first point to consider is that the residual solid material left in the upper mantle has had most of its low-temperature components removed by partial melting. Thus the residual material consists predominantly of the higher-temperature minerals in the original peridotite, and it follows that the chemistry of any liquid produced will be very different from that of common basalts — it will contain much more olivine, for example. In effect, therefore, suitable basaltic liquid from which to generate andesite or granite by fractional crystallization can only be produced *once* from any segment of upper mantle peridotite.

The second point to note is the relative volumes of material shown in Figure 4.22. Clearly, there is insufficient volume of material in the mantle wedge to generate the continental crust above it simply by fractional crystallization of a basaltic magma. Moreover, the large volumes of cumulates required by this model have not been found. These points taken together suggest that fractional crystallization may not be the only mechanism responsible for generating magmas of intermediate and acid compositions at destructive margins. We must therefore consider other models for the generation of these rock types. Perhaps they can be derived by partial melting in the *crust*, as well as by fractional crystallization of magmas that were derived by partial melting in the mantle. To explore this possibility, we must look at a new ternary system, quartz–albite–orthoclase.

4.3.2   THE QUARTZ–ALBITE–ORTHOCLASE SYSTEM

Since many granites are made up of over 90% of just three minerals, quartz, albite-rich plagioclase and orthoclase, it is possible to make quite accurate comparisons between the compositions of liquids within the synthetic quartz–albite–orthoclase (Qz–Ab–Or) system and natural 'granites'. Albite and orthoclase are the sodium and potassium end-members of the alkali feldspar system (Appendix 1, Block 3). In a sense, albite is the cornerstone of the feldspar minerals. On one side, it exists in complete solid solution with anorthite in the plagioclase feldspars, and on the other, it also undergoes *partial* solid solution with orthoclase under common geological conditions, these two together making up the alkali feldspars. Mixtures in the binary quartz–albite and quartz–orthoclase systems exhibit simple eutectic behaviour similar in some ways to that shown in Figure 4.8 for An and Di. The relationships between albite and orthoclase, however, are more complex. The system of these two feldspars shows both partial solid solution and a temperature minimum, the characteristics of which are explained below.

Figure 4.23 is a phase diagram of the alkali feldspar system under fixed pressure conditions. The general principles you have already studied also apply to the crystallization relationships between albite and orthoclase. Their system is like two separate plagioclase systems joined together at a common low-temperature point. This point has a composition of about Or_{30} (70% albite, 30% orthoclase). Solid solution takes place on both sides of this minimum point (D in Figure 4.23) in just the same way as it does in the pure plagioclase system (Figure 4.5), although at lower temperatures ($< 700 \, °C$), only partial solid solution occurs between albite and orthoclase as discussed below.

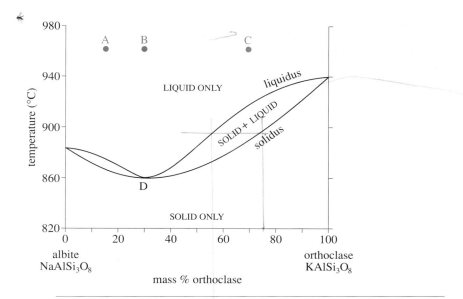

Figure 4.23 The alkali feldspar binary system showing a temperature minimum (D), determined at a pressure of $10^8\,N\,m^{-2}$. For A, B and C, see ITQ 13.

ITQ 13

(a) What is the composition of the first crystals to form on cooling melts of compositions A, B and C in Figure 4.23?

(b) Assuming that bulk equilibrium is maintained between the crystals and the liquid throughout crystallization, what is the composition of the last crystals to form just prior to complete solidification of samples A, B and C in Figure 4.23?

Contrast your answer to ITQ 13(b) with what happens if fractional crystallization takes place in this system. As more and more crystals are removed, *all* liquids, *regardless of their starting composition*, move towards the minimum point D. This is arguably the most significant feature of the alkali feldspar system.

The minimum point D depicts the composition of the lowest temperature liquid that can exist in the Ab–Or system at this particular pressure. That is a feature it shares with a eutectic point in a binary eutectic system — but can you see what makes a temperature minimum *different* from a eutectic point?

❏ How many minerals are in equilibrium with the liquid at the eutectic point in a binary eutectic system such as Di–An (Figure 4.8)?

■ The answer is two. The eutectic point is the only place in a binary eutectic system where *both* minerals (in that case, diopside and anorthite) can coexist with the liquid.

❏ What is the composition of the mineral(s) in equilibrium with the liquid at the minimum point D in Figure 4.23?

■ Crystals in equilibrium with a liquid plot on the solidus at the same temperature as that of the liquid. At the minimum point D in Figure 4.23, they therefore have the *same composition as the liquid*.

Therefore, during equilibrium crystallization, all crystals at a temperature minimum have the same composition, and thus, unlike the situation at the eutectic point, only *one* mineral can be in equilibrium with the liquid at a binary temperature minimum. All crystals at D have composition Or_{30}. So at a temperature minimum in a binary system, one mineral coexists with the liquid that has the same composition as that of the liquid. In contrast, at a eutectic point in a binary system, two minerals (with distinct composition) coexist with the liquid.

An additional complication of the alkali feldspar system (not shown by Figure 4.23) is that for temperatures below about 700 °C, alkali feldspar crystals near the middle of the Ab–Or range may separate into two distinct compositions, one rich in Or and the other in Ab. This has been referred to before as partial solid solution, and is due primarily to the disparity in ionic radius between K^+ and Na^+. They differ by about 35% and thus contravene the general rule introduced in Block 3 (Section 3.2.2) that one ion may replace another completely to form a solid-solution series if their size difference does not exceed 15% of the radius of the smaller ion. In the alkali feldspars, K^+ and Na^+ will substitute for one another and form a complete solid-solution series at high temperatures. If these crystals cool slowly, element diffusion will occur between two distinct crystal structures, one of which accommodates Na^+, and the other K^+. This process is called **exsolution** and is the result of one solid crystal separating into two minerals of distinct compositions. In the alkali feldspars, it results in complex intergrowths between albite-rich and orthoclase-rich crystals (Figure 4.24).

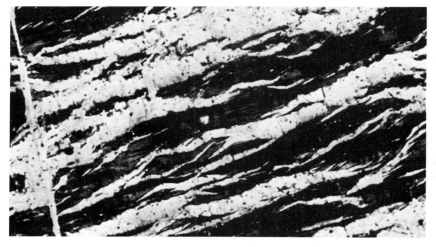

Figure 4.24 Photomicrograph of an alkali-feldspar intergrowth. Dark areas, albite; light areas, orthoclase. Each phase is homogeneous. (Magnification ×40.)

If we now put the three binary systems (quartz–albite, quartz–orthoclase and orthoclase–albite) together into a three-component Qz–Ab–Or system, the shape of the liquidus surface in a three-dimensional diagram of composition against temperature looks like Model 3 in your Kit. Figure 4.25 is the phase diagram, (or temperature contour map) of that system. It consists of two portions, the field in which quartz crystallizes first and the field in which alkali feldspar crystallizes first.

The quartz–albite and quartz–orthoclase binary systems have *eutectic points* at A and B respectively; these are joined by a cotectic curve AMB in the ternary system. This curve has the same characteristics as the cotectic curve we discussed in the Di–Ab–An system (Figure 4.9) except that it slopes 'downhill' *towards M* from *both* the eutectic points A and B: M is the lowest point on the curve AMB and therefore the lowest temperature on the entire liquidus surface.

❑   Look at Model 3 from the side of the albite–orthoclase system. Can you suggest why the cotectic curve AMB has its 'basin-like' form and does not, for example, just slope 'downhill' continuously from A to B?

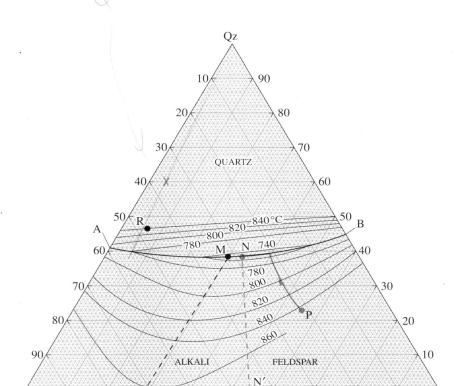

Figure 4.25 Diagram of the quartz–albite–orthoclase system determined at a pressure of $10^8\,N\,m^{-2}$. For clarity, the contours for Qz-rich and Or-rich compositions have been omitted. (For explanation of the letters, see text.)

■ The answer lies in the form of the liquidus curve in the binary alkali feldspar system albite–orthoclase — the cotectic curve AMB reflects (although at a lower temperature) the form of the liquidus curve in Figure 4.23. The dashed line DM (in Figure 4.25) marks the composition of the temperature minimum on the liquidus surface for alkali feldspar in the Qz–Ab–An system.

In summary, in the system Qz–Ab–Or, the quartz and alkali feldspar stability fields are separated by a curve (AMB) running between the quartz–albite (A) and the quartz–orthoclase (B) eutectics. This cotectic curve slopes downhill (down-temperature) from all compositions to a single point M. This is commonly called the **granite minimum**, and its composition is nearly equivalent to that of many naturally occurring granites. You will see in a moment why this should be so.

You should now listen to the third part of the audiocassette AV 08 'Phase Diagrams', which will introduce Model 3 in your Kit, the Qz–Ab–Or ternary system. It lasts about 4 minutes.

Let us now consider the crystallization path of a particular melt P (Figure 4.25) in the Qz–Ab–Or system.

❑ What is the temperature at which the first crystalline phase starts to separate from a melt of composition P?

■ 840–845 °C.

❑ What is the first crystalline phase to separate on cooling a melt of composition P?

■ P plots in the alkali feldspar field; thus the first crystals are of alkali feldspar.

❑ Will the first alkali feldspar crystals contain more or less orthoclase than the liquid from which they crystallized?

■ More. Qualitatively this can be seen by comparing the liquidus and solidus curves for orthoclase-rich compositions in Figure 4.23

❑ As the temperature falls, how will the composition of the melt change?

■ The melt (or liquid) is experiencing two effects:
(i) The feldspar that is crystallizing is richer in orthoclase than the liquid, and thus the remaining liquid must be getting richer in albite.
(ii) Because feldspar is crystallizing, the melt must be also getting richer in quartz. The net result is that the melt moves along a curved path towards the cotectic curve AMB.

❑ What happens when the melt reaches the cotectic curve?

■ Quartz will start to crystallize, and, with continued crystallization of feldspar and quartz, the melt composition will move along the cotectic towards the minimum M.

❑ Assuming that complete equilibrium is maintained between the crystals and the liquid, what is happening to the feldspar composition in the mean time?

■ It is changing by reaction with the melt, and its composition becomes richer in albite as the melt moves *towards* the minimum M. In principle, this is the same as equilibrium crystallization in the binary Ab–Or system (Figure 4.23). However, the temperature of the minimum in that system (D) is higher than that in the ternary Qz–Ab–Or system (point M in Figure 4.25).

❑ As the temperature falls and the liquid moves along the cotectic curve, what will be the composition of the last crystals to form before solidification is complete?

■ The last crystals to form will be quartz and alkali feldspar, which has the same composition as the feldspar in the original liquid (about Or_{70}). (Again this assumes that crystallization took place under equilibrium conditions.) The liquid in equilibrium with these last crystals lies on the cotectic curve AMB at about point N (Figure 4.25). The position of N can be determined from combining Figures 4.23 and 4.25. The starting composition (P) is approximately Qz = 23%, Ab = 22%, Or = 55%. The orthoclase content of the feldspar is therefore $(100 \times 55)/(55 + 22)\% = Or_{71}$. From Figure 4.23, a solid alkali feldspar with approximate composition Or_{70} is in equilibrium with a liquid with composition Or_{54}. If we plot this on the alkali feldspar face of the ternary system (N′ on Figure 4.25), N will lie at the point on the cotectic curve, AMB, that intersects the line Qz–N′.

❑ In the absence of fractional crystallization, what will be the bulk composition of the final product?

■ It will have the same overall composition as the melt we started with, P.

❑ If there had been fractional crystallization, what would then be the composition of the final liquid?

■ Fractional crystallization drives all liquids towards the composition of the granite minimum M.

ITQ 14

Determine the crystallization path and crystallization histories of a liquid of composition R (Figure 4.25), both (a) under equilibrium, and (b) under non-equilibrium conditions.

You will, we hope, now appreciate that phase relations in the synthetic granite system suggest that, under conditions of fractional crystallization, liquid containing quartz and alkali feldspar will always finish crystallizing in a rather restricted area of the system near the granite minimum. If naturally occurring granites originated by fractional crystallization, their compositions should lie near to that of the granite minimum. To assess this hypothesis, the composition of a large number of natural granites have been plotted on the phase diagram. The results are shown in Figure 4.26. There is a striking concentration of points within a small area close to the granite minimum which suggest that these granites *could* have originated by processes of fractional crystallization of silicate liquids. However, there is another process which could also explain why many granites lie so close to this minimum.

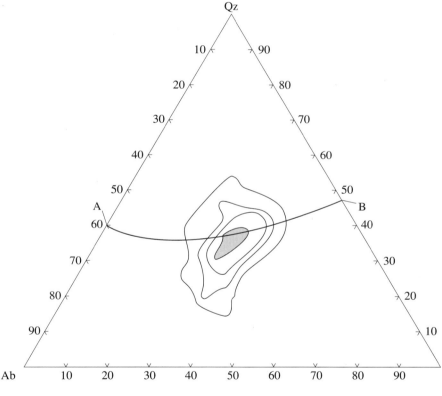

Figure 4.26 Chemical compositions of over 500 granitic rocks in terms of the proportions of quartz, albite and orthoclase. The results have been contoured on the basis of frequency and 90% of the analyses fall in the shaded area in the centre. The curve AB is the cotectic curve for the Qz–Ab–Or system.

❑ Can you recall what this might be?

■ It is the process of partial melting; as a solid is heated, the *first* part to melt is the lowest-temperature melting fraction, which is the *last* part to solidify during fractional crystallization. In the case of granites, we may be more specific: fractional crystallization of *any* mixture of alkali feldspars and quartz leads to a final composition close to the granite minimum. Conversely, if *any* mixture of alkali feldspars and quartz is heated until it begins to melt, the composition of the first formed melt will lie on the cotectic curve close to the granite minimum.

Partial melting will not result in a first melt of *precisely* the minimum-melt composition unless the albite/orthoclase ratio of the unmelted rock happens to be that of composition D in Figure 4.23. For starting compositions more orthoclase-rich than D, the partial melt will be albite-enriched, and for starting compositions more albite-rich than D, the partial melt will be orthoclase enriched. Since all melts will lie on the ternary cotectic curve (AMB, Figure 4.25), partial melting of a wide range of rocks containing quartz and alkali feldspar will result in compositions much closer to the granite minimum than the composition of the original rock.

Thus, both partial melting and fractional crystallization processes may account for the abundance of naturally occurring igneous rocks with compositions around the granite minimum.

Many rocks of intermediate composition (andesites and dacites) contain significant amounts of quartz, alkali feldspar and plagioclase. Thus, from our discussions of the phase relations in the synthetic Qz–Ab–Or system, we would predict that partial melting of rocks with intermediate SiO_2 contents should produce liquids with compositions near the granite minimum as observed in Figure 4.26. Geologists now accept that granitic rocks are generated by different processes under different circumstances. Some reflect fractional crystallization of basic parental magmas derived by partial melting within the upper mantle, others are partial melts of existing continental rocks. There is also a third possibility, which is that mantle-derived magmas may be 'contaminated' by the melting or partial melting of continental material as they pass through the crust. All three processes result in rocks with very similar major element compositions (that is, at or near the granite minimum), and thus it is impossible to distinguish the relative roles of each process solely on the basis of major element analyses.

4.3.3   TRACE ELEMENT BEHAVIOUR DURING GRANITE FORMATION

We know from Block 3 and Section 4.2.6 that the concentrations of trace elements in a crystallizing magma are controlled by both the initial concentration in the magma and by the fractional crystallization of minerals from that magma. The relationship between the ratio of the concentration of a trace element in the liquid, relative to that present in the source (C_1/C_o), the proportion of melt remaining (F), and the bulk partition coefficient (D) is shown graphically in Figure 4.27 (calculated from Equation 4.4).

It can be seen from Figure 4.27 that the concentration of a trace element in a magma will only be unaffected by fractional crystallization if the bulk partition coefficient (D) of the crystallizing minerals equals 1. For incompatible elements ($D < 1$), the trace-element concentration in the liquid will increase, and for compatible elements ($D > 1$), the trace-element concentration in the liquid will decrease with decreasing F.

We can draw a similar series of curves that describe the behaviour of trace elements with varying bulk partition coefficients under conditions of partial melting (Figure 4.28), based on the partial melting Equation 4.3. Note that for these curves, F represents the proportion of melt relative to the original rock.

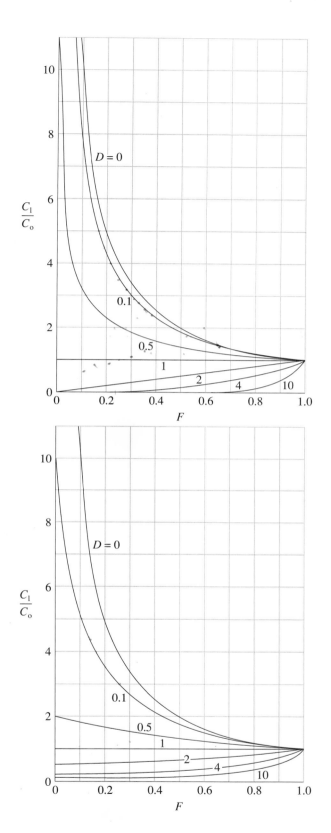

Figure 4.27   Trace element behaviour during fractional crystallization. C_l/C_o (the concentration of a trace element in the residual liquid C_l, divided by that in the original liquid C_o) is plotted against proportion of original liquid remaining (F) for different bulk partition coefficients (D).

Figure 4.28   Trace element behaviour during partial melting. C_l/C_o (the concentration of a trace element in the magma divided by that in the source) is plotted against proportion of melt formed (F) for different bulk partition coefficients (D).

You should notice that partial melting will also result in an increase in the concentration of incompatible elements in the melt, particularly for small melt fractions ($F < 0.2$). The melt will be depleted in compatible elements, and for large bulk partition coefficients ($D > 4$) the melt will contain very low concentrations even for quite high melt fractions.

In general, we can conclude that dramatic increases in the concentrations of incompatible elements can be obtained either by extreme fractional crystallization or by very small melt fractions during partial melting.

We shall now use both sets of curves to see whether trace-element concentrations in granites (Table 4.6) can determine whether granite magmas result from fractional crystallization of a basic magma or partial melting of the continental crust.

The first point to notice about Table 4.6 is that Rb, with a low bulk partition coefficient, is an incompatible element whereas Sr is compatible. In other words, during melting or crystallization, Rb will concentrate in the magma whereas Sr will be more strongly partitioned into minerals, such as feldspar, that coexist with the liquid.

Table 4.6   Rb and Sr concentrations (in ppm) for typical andesite and granite compositions from the Andes, with bulk partition coefficients (D).

	Andesite	Granite	D
Rb	70	140	0.1
Sr	340	100	3.5

ITQ 15

Using Figure 4.27 and the data in Table 4.6, determine the proportion of melt left if a liquid of andesitic composition undergoes fractional crystallization to form a liquid of granitic composition based on (i) Rb and (ii) Sr.

Depending on the trace element we examine, somewhere between 40 and 60% of the original andesitic magma is required to be removed as cumulates by fractional crystallization to generate the observed Rb and Sr concentrations in the granite. One reason why the value for F differs according to the trace element used is that the assumed value of D, the bulk partition coefficient, may be inaccurate, and indeed may vary during fractional crystallization, partly because the mineralogy of the crystal-lizing phases changes and partly because the mineral/liquid partition coefficients change with the composition of the liquid.

ITQ 16

Using Figure 4.28 and the data in Table 4.6, what proportion of the andesite would be required to melt to generate the granite based on (i) Rb and (ii) Sr trace-element concentrations?

The estimates of the melt fraction required to form the granite from melting the andesite therefore varies widely from 44% (based on Rb) to 4% (based on Sr). This is probably because the equation on which the model is based is unrealistic. We have assumed a constant D during the melting event, but minerals can be used up in the source, or as we see in Section 4.4, new minerals can be formed by the melting reaction. In either case, D is not strictly a constant. We shall also learn from Section 4.4 that metasediments rather than igneous rocks are likely to provide source rocks for granites that have formed from partial melting of the crust. All these variables can be modelled quantitatively, but more elaborate models than are presented in Figures 4.27 and 4.28 are beyond the scope of the Course.

However, we can make a general conclusion from considering simple quantitative modelling of the Rb and Sr contents of granites and of magmas of intermediate compositions as presented in Table 4.6. *Either* partial melting *or* fractional crystallization can explain the high concentrations of incompatible elements like Rb, or low concentrations of compatible elements like Sr. Trace elements, which have proved so powerful in identifying the conditions of partial melting in the mantle (where the mineralogy of the source is well known) and in tracing the crystallizing minerals from a suite of rocks related by fractional crystallization (where the starting magmatic composition is known), are of little value in determining the origin of granites, for in this case we know neither the process nor the source composition. In short, we have too many variables and not enough equations.

Table 4.6 also tells us that the Rb/Sr ratio of granite is higher than that of andesite. In general, Rb/Sr for basalts <0.1, for andesites ~0.2, and for granites >1. We shall see in the following sections that the high Rb/Sr ratios found in granitic rocks, relative to those of more basic rocks, provide a basis for isotope geochemists to determine not only how individual granites have formed but also how rapidly the continental crust has been extracted from the mantle through geological time.

4.3.4 RADIOACTIVITY APPLIED TO DATING

Atoms of the same element which have different masses are called isotopes. Some isotopes are unstable — they decay, and in so doing emit radiation. This process of radioactive decay has been harnessed by geologists both as an isotopic clock to date events throughout most of Earth history and as a sensitive tool for the study of the chemical evolution of the planet.

In this Section, we shall outline the principles of radioactive decay and show how they may be applied to the dating of rocks and minerals.

> The basic *law of radioactive decay* states that the rate at which the number of radioactive atoms decreases by decay is proportional to the number of such atoms present: the rate of decay is *exponential*.

During radioactive decay, the 'parent' atoms are decaying (and emitting radiation) and thus their numbers *decrease* exponentially with time. If P_0 is the number of original parent atoms and the number of those surviving ('daughter' atoms) after time t is called P, the decay process may be written as

$$P = P_0(1/2)^n \qquad \text{(Equation 4.5)}$$

where n is the number of half-lives of the decay scheme concerned, the **half-life** (τ) being the time taken for half the atoms present at any given time to decay.

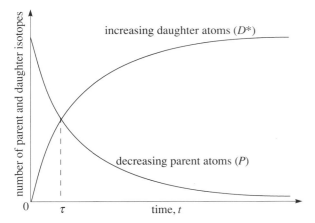

Figure 4.29 The changing number of parent and daughter atoms during radioactive decay. Note that growth in the number of daughter atoms (D^*) is the mirror image of the decay curve of the number of parent atoms (P).

Figure 4.29 illustrates how the number of parent atoms· (P) decreases with time. Note that, as we stated above, the rate at which P decreases depends on the value of P. Thus when P is 10, it is reduced by 5 in the next unit of time, and yet when P is 2, it is reduced by only 1 in the *same* length of time. This is the nature of an *exponential* decay process. We may also assume that the decay of *one* radioactive parent atom produces *one* daughter atom; thus, as illustrated in Figure 4.29, the number of daughter atoms produced by radioactive decay increases as rapidly as the number of parent atoms decreases.

If a number of parent atoms decays to a number of daughter atoms $D*$ over a period of time t, then $D*$ is related to the number of parent atoms remaining (P) by

$$D* = P(e^{\lambda t} - 1) \qquad \text{(Equation 4.6)}$$

where λ is a constant, termed the **decay constant**, which has a characteristic value for each species of radioactive isotope.

In Figure 4.29, the half-life (τ) is the time at which $P = D*$. At this time, $t = \tau$, so that $1 = (e^{\lambda \tau} - 1)$ (from Equation 4.6) or $e^{\lambda \tau} = 2$.

Taking logarithms to the base e (written ln), $\lambda \tau = \ln 2$, or 0.693. Thus, the half-life

$$\tau = \frac{0.693}{\lambda}. \qquad \text{(Equation 4.7)}$$

In isotopic dating, we usually refer to the decay constant, rather than the half-life, of an isotopic system, but as you can see from Equation 4.7, these two constants are simply related to each other.

Returning to the decay equation (4.6), the term ($e^{\lambda t} - 1$) may be approximated to λt. This is because geologists choose decay schemes with half-lives much longer than the ages of geological events that they are dating. Thus, on Figure 4.29, we are only concerned with the section of the decay curve to the left of τ ($t < \tau$), which is close to a straight line.

Thus, we may rewrite Equation 4.6 in an approximate form:

$$D* = P \lambda t. \qquad \text{(Equation 4.8)}$$

We stated that, in this equation, $D*$ is the number of daughter atoms produced by the decay of parent atoms *over a period of time t*. However, if we determine the total number of daughter atoms (D) at the present time, only some of them will have been produced during the time period that we are trying to date — some may have been present before.

Thus, the total number of daughter atoms (D) consists of those produced by radioactive decay during the period of time t ($D*$), plus those already present at the time of formation (D_0); that is, $D = D_0 + D*$.

Since $D* = P \lambda t$ (Equation 4.8),

$$D = D_0 + P \lambda t. \qquad \text{(Equation 4.9)}$$

This is the equation of a straight line and is the general equation used to calculate the ages of rocks and minerals from radioactive decay schemes. Note that geological ages are expressed in numbers of years *before present* and the units are usually given in millions of years (Ma). Thus, if a rock is described as 500 Ma old, it means that 500 million years have passed since it was formed.

4.3.5   THE DATING OF ROCKS AND MINERALS BY Rb–Sr

Geologists use several isotopic systems for dating geological events, including the decay of potassium to argon and of uranium to lead. However the most widely used technique is based on the decay of rubidium to strontium which not only can date the event, but also can provide information on the source of the rock.

Rubidium occurs in two isotopic states $^{87}_{37}\text{Rb}$, and $^{85}_{37}\text{Rb}$ (Table 4.7). One of these isotopes ($^{87}_{37}\text{Rb}$) is unstable and decays by β-radiation (effectively the transformation of a neutron to a proton plus an electron which is expelled as a negatively charged β-particle) to produce one of the four stable isotopes of strontium:

$$^{87}_{37}\text{Rb} \xrightarrow{\text{β-particle decay}} {}^{87}_{38}\text{Sr}$$

Table 4.7 Present-day abundances of Rb and Sr isotopes

Isotope	$^{87}_{37}\text{Rb}$	$^{85}_{37}\text{Rb}$	$^{88}_{38}\text{Sr}$	$^{87}_{38}\text{Sr}$	$^{86}_{38}\text{Sr}$	$^{84}_{38}\text{Sr}$
Abundance (% of element)	28*	72	83	7†	10	0.5

* unstable isotope   † increased by decay of $^{87}_{37}\text{Rb}$

The decay of radioactive $^{87}_{37}\text{Rb}$ to stable $^{87}_{38}\text{Sr}$ clearly results in changes in the isotopic composition of Rb and Sr. Isotopes are measured by mass spectrometry, and since mass spectrometers measure *ratios* of isotopes, an isotope composition is expressed as a ratio of two isotopes, for example $^{87}\text{Sr}/^{86}\text{Sr}$. (Note that it is both convenient and conventional not to include the atomic number when writing these isotopes.)

ITQ 17

(a)   With the passage of time, will the following ratios in a sample containing both rubidium and strontium increase, decrease, or remain unchanged?

   (i) $^{87}\text{Rb}/^{85}\text{Rb}$;   (ii) $^{87}\text{Sr}/^{86}\text{Sr}$;   (iii) $^{86}\text{Sr}/^{84}\text{Sr}$.

(b)   In a rock that contains no Rb at all, will the $^{87}\text{Sr}/^{86}\text{Sr}$ ratio increase, decrease, or remain unchanged with the passage of time?

(c)   In a rock that initially contains no Sr at all, will the $^{87}\text{Rb}/^{85}\text{Rb}$ ratio increase, decrease, or remain unchanged with the passage of time?

Since almost all rocks and minerals contain at least small amounts of Rb and Sr, you can now see why the isotope abundances given in Table 4.7 are for present-day Rb and Sr.

Clearly, with time, the $^{87}\text{Sr}/^{86}\text{Sr}$ ratio of any rock that contains Rb will increase and its $^{87}\text{Rb}/^{86}\text{Sr}$ ratio will decrease, because of the production of ^{87}Sr from ^{87}Rb. The degree of enrichment of ^{87}Sr in a rock therefore depends both on how old it is, and on the amount of rubidium relative to strontium that it contains, that is, its Rb/Sr ratio. This is the basis of the Rb-Sr method for dating rocks and minerals.

You should now listen to audiocassette AV 09 'Isotope Geochemistry and Geochronology' as an introduction to the subject. The audiocassette lasts about 15 minutes.

For Rb and Sr, ^{87}Rb decays to ^{87}Sr, and thus ^{87}Rb is the parent atom and ^{87}Sr is the daughter atom. The present-day ^{87}Sr abundance results from the addition of the number of atoms produced by radioactive decay during the period of time t ($^{87}\text{Sr}^*$) to the number already present ($^{87}\text{Sr}_0$) at the time of formation of the sample.

From Equation 4.8, $^{87}\text{Sr}^* = {}^{87}\text{Rb}\lambda t$.

Substituting in Equation 4.9, we get $^{87}\text{Sr} = {}^{87}\text{Sr}_0 + {}^{87}\text{Rb}\lambda t$.

However, as mentioned above, mass spectrometers determine *ratios* of isotopes, and thus it is convenient to divide this equation across by a stable isotope that is not involved in the decay scheme, in this case ^{86}Sr.

$$\left(\frac{^{87}\text{Sr}}{^{86}\text{Sr}}\right) = \left(\frac{^{87}\text{Sr}}{^{86}\text{Sr}}\right)_0 + \left(\frac{^{87}\text{Rb}}{^{86}\text{Sr}}\right)\lambda t \qquad \text{(Equation 4.10)}$$

The present-day Sr-isotope composition $^{87}Sr/^{86}Sr$ can be determined by a mass spectrometer, and the $^{87}Rb/^{86}Sr$ ratio can be calculated once the present-day concentrations of Rb and Sr have been determined. Note that to measure the *element* concentrations of trace elements (or major elements) does not require a mass spectrometer (which measures isotopic ratios) but requires a much more routine procedure called X-ray fluorescence analysis.

Assuming that the decay constant λ has already been measured experimentally, then we may calculate the period of time t (the age before present) once we have estimated the Sr-isotope composition of our sample at the time of its formation, $(^{87}Sr/^{86}Sr)_o$.

At first glance, that may seem rather a tall order. Most rocks are many millions of years old and clearly none of us were around at the time when they were formed ($t = 0$) to measure the $^{87}Sr/^{86}Sr$ ratio of the sample at that time. However, you may have noticed that Equation 4.10 is in the form $y = c + mx$, in other words, it defines a straight line.

> For samples of the same age and same $(^{87}Sr/^{86}Sr)_o$, a graph of $^{87}Sr/^{86}Sr$ against $^{87}Rb/^{86}Sr$ should be a straight line with a slope of λt and an intercept corresponding to $(^{87}Sr/^{86}Sr)_o$.

Note also that since $(^{87}Sr/^{86}Sr)_o$ is the Sr-isotope ratio at the time of formation ($t = 0$), it is usually referred to as the **initial Sr-isotope ratio**.

Equation 4.10 can therefore lead to the determination of an age of a suite of rocks for which $^{87}Rb/^{86}Sr$ and $^{87}Sr/^{86}Sr$ ratios have been determined experimentally. As an example, let us consider four samples from a hypothetical suite of volcanic rocks that were erupted during the same volcanic event t million years ago. These samples were selected because (a) they have different Rb/Sr ratios, and (b) they are believed to have been derived from the *same* primary magma (by fractional crystallization of varying amounts of plagioclase and clinopyroxene). This is significant, because we may assume that rocks derived from the same magma will, at the time of their formation, have shared the same $^{87}Sr/^{86}Sr$ ratio, equivalent to that of the parental magma.

When the rocks formed, therefore, the processes of fractional crystallization ensured that the *trace-element ratio* (Rb/Sr) would be *different* in these rocks. In contrast, the *Sr-isotope ratio* ($^{87}Sr/^{86}Sr$) would be the *same*, because fractional crystallization does not alter the relative proportions of isotopes of a given element.

We can represent this graphically, as in Figure 4.30, which is a graph of $^{87}Sr/^{86}Sr$ against $^{87}Rb/^{86}Sr$. The line AB on that graph represents the situation when the rock samples showed the same $^{87}Sr/^{86}Sr$ ratios. In other words, the situation immediately after the rocks crystallized from the same isotopically homogeneous magma.The sloping line AC represents the situation today: the $^{87}Rb/^{86}Sr$ ratios are still different, but now so are the $^{87}Sr/^{86}Sr$ ratios.

❑   What has happened?

■ ^{87}Rb has decayed to ^{87}Sr over the time period t from the formation of these rocks to the present day. This has had two effects:
(i) it has *reduced* the amount of ^{87}Rb present, and so the $^{87}Rb/^{86}Sr$ ratios have decreased slightly:
(ii) it has *increased* the amount of ^{87}Sr, and so the $^{87}Sr/^{86}Sr$ ratios have increased.

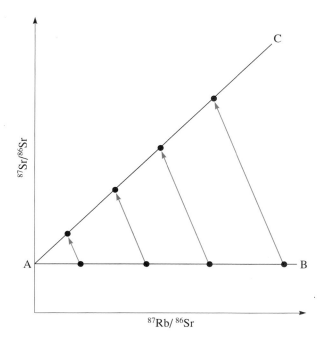

Figure 4.30   A graph of ^{87}Sr/^{86}Sr against ^{87}Rb/^{86}Sr (an isochron diagram) illustrating how four samples of the same age but different Rb/Sr ratios evolve from a horizontal line (AB) at the time of their formation, to plot on a straight line (AC) with slope equal to λt at the present time.

❑   But why have the ^{87}Sr/^{86}Sr ratios increased by different amounts along the line AC?

■   The amount of ^{87}Sr produced by radioactive decay depends on the amount of ^{87}Rb present. However, on Figure 4.30 we have plotted isotope ratios: hence the *relative* increase in ^{87}Sr (the increase in ^{87}Sr/^{86}Sr) depends on the *relative* amount of ^{87}Rb (or the ratio ^{87}Rb/^{86}Sr). After a given period of time, the samples with the highest ^{87}Rb/^{86}Sr ratios will also be those that generate the highest ^{87}Sr/^{86}Sr ratios by radioactive decay.

From Equation 4.10, we know that a graph of ^{87}Rb/^{86}Sr against ^{87}Sr/^{86}Sr gives a straight line with a slope equal to λt, where t is the time since the samples last showed the same isotopic composition, which for igneous samples indicates the time since they were last molten.

❑   Will older samples lie on lines with steeper or shallower slopes on diagrams such as Figure 4.30?

■   The steeper the slope, the larger the value of t, and hence the older the samples. The line AB in Figure 4.30 has zero slope and so t is zero. This is what we said at the beginning — the line AB represents the situation when the rocks first formed. The line AC represents the situation at the present day when the rocks have an age t. And in the future, when t increases as the rocks become older, the slope will increase further. From this example, you should see that groups of rocks with *different* ages give different slopes and therefore can be dated using this method.

The straight lines AB and AC on Figure 4.30 connect samples *of the same age*, and they are therefore called **isochrons**. The diagram of ^{87}Sr/^{86}Sr versus ^{87}Rb/^{86}Sr is an **isochron diagram**. The line AB is an isochron corresponding to an age of zero, while AC corresponds to an age of t. The initial Sr-isotope ratios can be read off the ^{87}Sr/^{86}Sr axis corresponding to ^{87}Rb/^{86}Sr = 0.

The samples in Figure 4.30 represent an idealized case. If we are to date samples by radioactive decay and if they are to plot on an isochron whose slope reflects their true age, then the following assumptions must be valid:

1   All the samples analysed must be of the same age. When samples are collected from the same igneous intrusion then clearly this is a reasonable assumption. However, if we collected samples over a large area where the geology was poorly understood, we might well collect samples of different ages without realizing it. In such a case, the isotope ratios we measured would *not* plot on a straight line.

2   All the samples must have the same initial Sr-isotope ratio when they are formed, $(^{87}Sr/^{86}Sr)_0$, otherwise they will not plot on the same straight line on an isochron diagram (AB on Figure 4.30). This assumption is most likely to be valid for a suite of igneous rocks thought to have crystallized from similar liquids; it is least likely in sediments composed of different fragments of existing material.

3   The third assumption is that between the formation of the samples and the present day, the $^{87}Sr/^{86}Sr$ and $^{87}Rb/^{86}Sr$ ratios have changed *only* by the process of radioactive decay; no Rb or Sr has been added to, or lost from, the system. This means that the isochron diagram is of no value if the Rb or Sr concentrations in the samples have been affected by fluids after their initial isotopic homogenization.

In practice, if any of these assumptions is not valid for a suite of samples, then the results will tend to scatter when plotted on an isochron diagram. Conversely, therefore, if all the results do plot on, or close to, a single straight line (for example line AC in Figure 4.30), it is probable that the assumptions have been upheld and that the samples really were formed at the calculated time *t*. It is one of the strengths of the isochron dating technique that the underlying assumptions are rigorously tested by the data.

Let us now see how these assumptions, and the method of determining ages from an isochron diagram, may be applied to a suite of natural rocks. Figure 4.31 illustrates results on metamorphosed granites from western Greenland plotted on an isochron diagram of $^{87}Sr/^{86}Sr$ against $^{87}Rb/^{86}Sr$. They fall on or close to a straight line that has been fitted to the data allowing for the uncertainty in the measurement of each isotopic ratio.

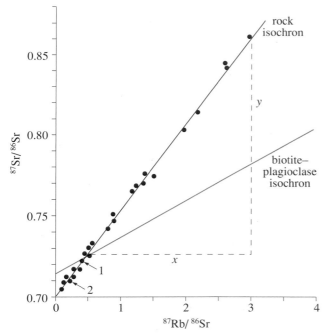

Figure 4.31   An isochron diagram presenting analyses of metamorphosed granites from western Greenland. The slope of the isochron = y/x.

The slope of the isochron on Figure 4.31 is given by y/x. The values of y and x may be read off their respective axes, $^{87}Sr/^{86}Sr$ and $^{87}Rb/^{86}Sr$:

$$y = 0.859 - 0.727 = 0.132$$
$$x = 3.0 - 0.5 = 2.5$$

and therefore

$$\text{the slope } \frac{y}{x} = \frac{0.132}{2.5} = 0.052\,8.$$

From the Equation 4.10 derived earlier, the slope of an isochron equals λt. The decay constant (λ) for the decay of ^{87}Rb is $1.42 \times 10^{-11}\,a^{-1}$. (This value is directly related to the half-life of ^{87}Rb, which can be calculated to be $48\,802$ Ma from Equation 4.7.) Therefore,

$$0.052\,8 = 1.42 \times 10^{-11}\,a^{-1} \times t$$

$$t = \frac{0.052\,8}{1.42} \times 10^{11} \text{ years}$$

$$= 3.7 \times 10^{9} \text{ years.}$$

Geological ages quoted in the scientific literature are always given uncertainties. The size of the uncertainty is determined from the errors in the isotopic measurements and the spread of the data from the straight line of the isochron diagram. The age of these rocks in western Greenland is given as $3\,700 \pm 50$ Ma. By convention, the quoted error is two standard deviations from the mean. In this case, there is a 95% probability that the true age lies between $3\,650$ and $3\,750$ Ma. In fact, these granites are among the oldest rocks known to be preserved on Earth.

❑ From Figure 4.31, can you determine the initial Sr-isotope ratio $(^{87}\text{Sr}/^{86}\text{Sr})_o$ of these western Greenland rocks?

■ The answer is the value of ^{87}Sr/^{86}Sr on the isochron where ^{87}Rb/^{86}Sr $= 0$, which can be read off as 0.700. The value calculated from the data is $0.700\,1 \pm 0.001\,0$.

ITQ 18

The blue line in Figure 4.31 is an isochron drawn through the isotope results for biotite mica and the plagioclase feldspar which were separated from one of the metamorphosed granites from western Greenland and then analysed. Assume that the slope of this isochron corresponds to the age of those minerals.

(a)   Are these minerals older, younger, or of the same age as the rocks whose age we have just calculated to be $3\,700$ Ma? (You should try to answer this part of the question without doing any calculation — simply compare the slopes of the two isochrons on Figure 4.31.)

(b)   Calculate the slope and hence the age of the isochron for biotite and plagioclase feldspar ($\lambda = 1.42 \times 10^{-11}\,a^{-1}$).

The answer to ITQ 18 illustrates a further point of which you should be aware: *minerals need not be of the same age as their host rock*. In the simplest case, an igneous rock crystallizes and clearly those minerals are of the same age as the rock itself. However, if at some later date the igneous rock is reheated and metamorphosed then new metamorphic minerals will grow. Their age will be that of the metamorphism and will be younger than that of the host rock. The results from western Greenland indicate that while granites were formed $3\,700 \pm 50$ Ma ago, they were then reheated so that metamorphic minerals were formed at about $1\,600 \pm 100$ Ma. Moreover, by analysing both the rocks and

minerals from different groups of rocks in this area, it has been possible
to document a fairly complex sequence of geological events.

4.3.6 INITIAL Sr-ISOTOPE RATIOS

The previous Section concentrated on how to determine the age of a rock
or a mineral. However, much useful information is also preserved in their
initial Sr-isotope ratios. What determines the variation in $(^{87}Sr/^{86}Sr)_0$
among suites of igneous rocks, and why do such ratios tend to be higher in
younger rocks?

> The value of an initial Sr-isotope ratio reflects both the chemistry (as
> mirrored in its Rb/Sr ratio) and age of the source region from which
> the particular rock was derived.

Let us return to our equation for determining the age of a sample from
the decay of ^{87}Rb to ^{87}Sr (Equation 4.10). We have established that this is
in the form of $y = mx + c$, and if we plot a diagram of $^{87}Sr/^{86}Sr$ against
$^{87}Rb/^{86}Sr$, we should get a straight line with a slope equal to λt.
However, we can also plot a second useful diagram in which the slope of
a straight line will no longer be λt, but rather $\lambda(^{87}Rb/^{86}Sr)$.

A graph of $^{87}Sr/^{86}Sr$ (y) against time t (x) will also yield a straight line
($y = mx + c$), but in this case the slope (m) is $\lambda(^{87}Rb/^{86}Sr)$. Figure 4.32 is
such a diagram, and its great advantage is that it enables us to illustrate
how $^{87}Sr/^{86}Sr$ ratios *change* with time. It is therefore called an **isotope
evolution diagram.** Note that for isotope evolution diagrams, time is
measured before present, and the present-day isotopic ratio (for $t = 0$) is
plotted on the right-hand side of the graph.

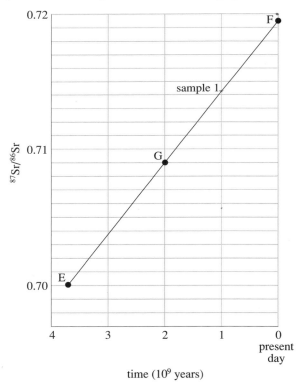

Figure 4.32   An Sr-isotope evolution
diagram illustrating the change in the
$^{87}Sr/^{86}Sr$ ratio of sample 1 (Figure
4.31) from 3 700 Ma (E) to the
present day (F).

Consider sample 1 from western Greenland (Figure 4.31). Its present-day
$^{87}Sr/^{86}Sr$ ratio was measured as 0.7195 ± 0.0002, and its initial Sr-
isotope ratio (that is, 3 700 Ma ago) was found to be 0.700 1 from the
isochron intercept on Figure 4.31. We may plot these two $^{87}Sr/^{86}Sr$ ratios
on the isotope evolution diagram (Figure 4.32) at 0 Ma (point F) and

3 700 Ma (point E), respectively. They are joined by a straight line EF whose slope is $\lambda(^{87}Rb/^{86}Sr)$. Moreover, since $^{87}Rb/^{86}Sr$ is approximately proportional to Rb/Sr (for most rocks $^{87}Rb/^{86}Sr \approx 2.9 \times Rb/Sr$), *the slope is also approximately proportional to Rb/Sr* — that is, the higher the Rb/Sr, the steeper the slope.

Equally important, however, is the fact that the line EF is the *path* along which this sample evolved from 3 700 Ma to the present day. Thus, if we wish to roll back the clock say 2 000 Ma, we may use the line EF to read off what the Sr-isotope ratio of this sample was at that time. 2 000 Ma ago it was at point G, and its $^{87}Sr/^{86}Sr$ ratio was therefore 0.709 0.

ITQ 19

A second sample (number 2 in Figure 4.31) of the 3 700 Ma old granites from western Greenland has a present-day $^{87}Sr/^{86}Sr$ ratio of 0.708 0 ± 0.000 1. Plot it on Figure 4.32, and draw the evolution path of this sample from 3 700 Ma to the present day. What was its Sr-isotope ratio 2 000 Ma ago?

These calculations are all very well, but why should we be bothering to find out about the Sr-isotope ratios of rocks at different times since they were formed? If the history of the rocks since their formation has been uneventful, then indeed there is little point in such calculations. But suppose, for example, that there had been powerful Earth movement and reheating of the western Greenland rocks 2 000 Ma ago, resulting in partial melting and the generation of granitic magmas.

❑   What would be the $^{87}Sr/^{86}Sr$ ratio of a granitic liquid produced by partial melting of sample 1 that occurred 2 000 Ma ago?

■   Assuming that equilibrium was maintained and the Sr-isotope ratio of the melt was therefore the same as that of the source rock, then a melt of sample 1 formed 2 000 Ma ago would have an $^{87}Sr/^{86}Sr$ ratio of 0.709 2 (G in Figure 4.32). The initial Sr-isotope ratio of the melt would therefore be 0.709 0.

Moreover, since you have also shown that, at that time, the $^{87}Sr/^{86}Sr$ ratio of sample 2 was different (0.703 8, ITQ 19) we may distinguish (on the basis of their initial Sr-isotope ratios) between granites derived by remelting samples 1 or 2. These and similar arguments can be applied to resolve far-reaching geological problems as we shall see.

4.3.7   Sr ISOTOPES AND THE GENESIS OF GRANITES

In Section 4.3.1, we concluded that the intermediate and acid magmas could be derived either from the mantle (via fractional crystallization of more basic magmas) or by partial melting of crustal rocks. The first mechanism represents the formation of *new* continental crust, whereas in the second, existing crustal material is simply being remelted and redistributed. Since both may result in granitic end-products with the same major element, and even trace-element, composition, major and trace elements cannot be used to establish the origin of such rocks — we must turn instead to their isotope geochemistry and in particular to their initial Sr-isotope ratios. We shall illustrate this point by considering granite formation from the central Andes.

You know from Block 2 (Section 2.5.3) that the Andes lie on an active continental margin resulting from the eastwards subduction of the Nazca Plate beneath the South American Plate. Although granitic magmas have formed in abundance in this tectonic setting (Figure 4.21), not all granites result from the same process. We shall consider two contrasting examples.

The first granite is from the high mountains of Bolivia (B on Figure 4.33). The intrusion is one of a series of Bolivian granites that are associated with tin-rich veins that are exploited by the tin-mining industry of Bolivia. The second granite is from northern Chile and is part of a suite of diorites, granodiorites and granites that is typical of intrusions found along the Pacific coast of much of the Andes (C on Figure 4.33). None of the coastal intrusions is associated with mineralization. The oldest exposed rocks found in both regions are a suite of metamorphosed sediments which were deposited about 300 Ma ago.

Available knowledge about the tectonic setting of the Central Andes suggests that the granites could be derived by either (i) partial melting of the upper mantle, followed by fractional crystallization, or (ii) partial melting of metamorphosed 300 Ma old sediments. Let us first consider the evolution of $^{87}Sr/^{86}Sr$ of the upper mantle.

The Earth is about 4 550 Ma old, and it has many chemical similarities with chondritic meteorites. Thus, since the Earth formed at the same time and from the same materials as chondrites, it is generally believed to have had the same initial Sr-isotope ratio, 0.699 0 (point A, Figure 4.34). The Earth is believed to have an average Rb/Sr ratio of about 0.03, equivalent to $^{87}Rb/^{86}Sr = 0.09$, which determines the gradient of the evolution line for average or bulk Earth on an Sr-isotope diagram — line AB in Figure 4.34.

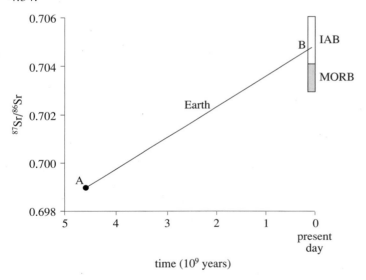

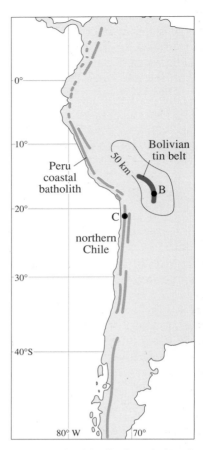

Figure 4.33 Distribution of plutonic rocks in the Andes (the Peruvian batholith shaded pale blue). B, Bolivian granite; C, northern Chile granite discussed in text. The contour line encloses region with crustal thickness > 50 km. The solid blue region marks the Bolivian tin belt.

Figure 4.34 Sr-isotope evolution diagram for the bulk Earth (line A–B). Also shown are the range of present-day $^{87}Sr/^{86}Sr$ ratios for IAB and MORB.

In addition to representing the evolution of the bulk Earth, the line AB depicts the evolution of any segment of the Earth in which the Rb/Sr ratio has remained unchanged from 0.03 for the last 4 550 Ma. (Strictly speaking, of course, Rb/Sr decreases with time as ^{87}Rb decays to ^{87}Sr, but since that causes the Rb/Sr ratio of the Earth to change by less than 2% since its formation, we may ignore it in this discussion.) Before we assume that this bulk Earth evolution line represents the upper mantle beneath a subduction zone, we should compare the present-day $^{87}Sr/^{86}Sr$ ratio of this simple model with that of modern basalts.

❑   In general, the $^{87}Sr/^{86}Sr$ ratios of the volcanic rocks from mid-ocean ridges (MORB) tend to be *lower* than that of the bulk Earth (as in Figure 4.34). Does that suggest that their upper mantle source areas have higher or lower Rb/Sr ratios than the bulk Earth?

■   Lower, since *lower* Rb/Sr ratios will with time result in *lower* $^{87}Sr/^{86}Sr$ ratios.

Rb tends to be more incompatible than Sr during the melting of peridotite, and thus particularly small volumes of partial melt tend to have slightly higher Rb/Sr ratios than their original source rocks.

❑ How, then, might upper mantle rocks come to have *lower* Rb/Sr ratios than that of the average Earth?

■ The simplest way of reducing the Rb/Sr ratio of a segment of mantle is to remove a small amount of (partial) melt from it. Moreover, since the Earth's crust is made up of material ultimately derived by partial melting in the mantle, many isotope geologists now believe that the low $^{87}Sr/^{86}Sr$ ratios of these upper mantle rocks reflect the extraction of partial melts to form the Earth's crust.

Now if we examine the Sr-isotope ratios of IAB, these are higher than those of MORB and on average are close to that of the average Earth (Figure 4.34). This is not perhaps what we would expect, because the mantle wedge at island arcs is repeatedly melted during the life of the arc, and therefore depletion of Rb/Sr should be more marked than at ocean ridges. The answer to this problem can be deduced from Figure 4.17. You may remember that in island-arc magmas both Rb and Sr are enriched by fluids derived from the subducted slab. However, Rb is more enriched than Sr, resulting in high Rb/Sr fluids. Hence, the resulting Rb/Sr ratio in the mantle wedge is increased by these fluids, which compensate for the depletion in Rb/Sr that is caused by melting. Our conclusion to all this is simple. The bulk Earth evolution line is a reasonable estimate for the evolution of the upper mantle, although locally, as beneath ocean ridges, the isotopic composition of the upper mantle may vary.

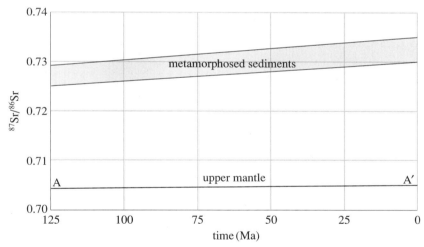

Figure 4.35 Sr-isotope evolution diagram for sediments from the Central Andes (for use with ITQ 20). Note that the apparently shallower slope for upper mantle evolution compared to Earth evolution in Figure 4.34 is a result of the expanded time-scale.

Typical samples of the metamorphosed sediments from the Central Andes have present-day $^{87}Sr/^{86}Sr$ ratios of 0.730 to 0.735. The Sr evolution lines are plotted on Figure 4.35 together with that of the upper mantle (taken from the bulk Earth of Figure 4.34). Note that the Rb/Sr ratios of the sediments are significantly higher than those in the mantle rocks, and that they therefore evolve along a line of slightly steeper slope on a Sr-isotope evolution diagram. They also form a band rather than a line, as there is some variation in $(^{87}Sr/^{86}Sr)_o$ within the metasediments.

❑ From our discussions of how Rb/Sr ratios in the liquid change during partial melting and fractional crystallization processes, why should these typical crustal rocks have higher Rb/Sr ratios than their mantle source rocks?

■ Ultimately, all sediments are the eroded fragments of igneous rocks, and continental sediments are predominantly derived from igneous rocks that are intermediate or acid in composition (>52% SiO_2). Rocks with higher SiO_2 contents are either produced by small degrees of partial melting, or by fractional crystallization from more basic magmas involving minerals that include plagioclase feldspar. In either case, the Rb/Sr ratio of the resultant granitic liquid will be higher than that in the original source rocks. Continental sediments have the high Rb/Sr ratios of the igneous rocks from which they are derived.

The trends in Figure 4.35 represent the best available estimate of how the Sr-isotope ratios of the two likely source regions for the Andean granites have evolved with time. If $^{87}Sr/^{86}Sr$ are determined for several samples from each granite, then their age and initial Sr-isotope ratios may be calculated using the isochron diagrams. We shall consider results from each granite in the following ITQ.

ITQ 20

The Chilean granite gave an age of 100 ± 5 Ma and an initial Sr-isotope ratio of $0.704\,5 \pm 0.000\,2$. The Bolivian granite by contrast is both younger (25 ± 1 Ma) and has a higher initial Sr ratio ($0.732\,5 \pm 0.000\,2$). Plot these results on Figure 4.35. What is the most likely source material for each granite?

Thus, Sr isotopes have allowed us to argue that an Andean granite that contains high concentrations of tin was derived from a different source region from one that does not. You may have noticed in Figure 4.33 that the Bolivian tin belt lies within the zone of unusually thick crust. This is no coincidence. Thickened crust results in steep geothermal gradients that trigger crustal melting as we shall see in Section 4.4. Moreover, a study of tin-bearing granites from all over the world suggests that the great majority, including those from southwest England, have high initial Sr-isotope ratios. An example of a biotite granite from Dartmoor that has been derived from crustal melting is provided in the Kit (Sample 7). Remelting sediments in the crust appears to be an effective method of concentrating certain elements, including tin (and incidentally, uranium) into the granitic magma.

The seemingly theoretical analysis of Sr isotopes has provided a possible tool for economic exploration. If we know where the Earth's continental rocks have been remelted, we might be able to identify zones of important mineralization associated with granite magmas.

You should now view video VB 06, 'Magmas in Scotland', which revises what you have learned about magma formation and examines the significance of Sr-isotope geochemistry for tracing the source of some Scottish granites. The video lasts about 27 minutes.

4.3.8   CRUSTAL GROWTH IN THE ANDES

Andesitic magmas with compositions close to that of average continental crust are formed at active continental margins. But how much of the voluminous magmatism generated at such margins represents new additions to the continental crust?

In the previous Section, we examined the isotope geochemistry of two Andean granites. One formed from the mantle, and was therefore an example of *crustal growth*. The second formed from melting sediments, and was therefore an example of *crustal remelting*. If we knew the initial Sr-isotope ratio of every igneous rock formed in the Andes, could we estimate how much crustal growth has occurred beneath the Andes?

Figure 4.35 might mislead us into thinking that this would be an easy task. Any granites that come from the mantle would lie on the upper mantle evolution line, and any that come from the crust would lie within the band for metamorphosed sediments. Unfortunately, available data from the Andes do not define two distinct groups but form a cloud that stretches from below the mantle evolution curve, to above the sediment evolution line. There are three reasons for this.

(i)   Not all crustal rocks will lie on the band shown for sediments in Figure 4.35. The band results from the isotopic characteristics of the source from which these particular sediments have been derived and the Rb/Sr ratio of the sediments. Other sediments, particularly older ones, would lie above the growth curve shown in Figure 4.35. Unfortunately, rocks from the lower crust have lower Rb/Sr ratios than do sediments and they form Sr evolution lines close to those of the upper mantle. So the initial Sr-isotope ratios of *all* granites formed from melting the crust would cover a wide range.

(ii)   Magmas may have more than one source. A magma may form in the mantle and be intruded into a magma chamber in the crust where fractional crystallization takes place (Figure 4.20). At the same time, some of the crustal rocks that form the walls of the chamber may be partially melted adding magma with a higher $^{87}Sr/^{86}Sr$ ratio. If this mixed magma is intruded into higher levels, it will have an initial Sr-isotope ratio that lies between that of a pure crustal source and that of a pure mantle source.

(iii)   The upper mantle evolution line shown is an average value, but the mantle is isotopically quite variable. In Figure 4.34, it can be seen that basalts from island arcs may have an $^{87}Sr/^{86}Sr$ ratio as high as 0.706 0, and even higher values have been recorded from some island arcs. So, strictly, the mantle evolves not along a line but within quite a wide band. There is considerable overlap between the isotopic ratios of some mantle rocks and rocks from the lower crust with low Rb/Sr ratios.

Because of the ambiguities in interpreting the source regions of many igneous rocks based on Sr isotopes alone, geochemists employ several isotopic systems each of which provides additional information about the identity of the magma source. Such techniques are beyond the scope of this Course. However, some useful patterns of Sr-isotope ratios emerge if we restrict our attention to the variations in the initial Sr-isotope ratios from a narrow belt of Andean rocks.

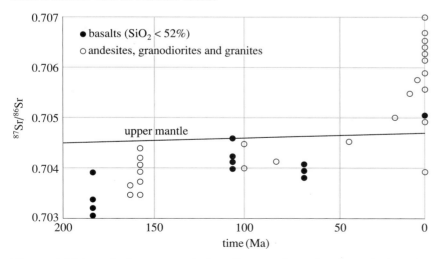

Figure 4.36   Sr-isotope evolution diagram for volcanic and plutonic rocks from northern Chile.

Figure 4.36 is a Sr-isotope evolution diagram for volcanic and plutonic rocks emplaced over 200 Ma from the coastal region of northern Chile. Basalts and more silica-rich magma compositions are distinguished from each other. The first point to notice is that many magmas fall below the mantle evolution line, particularly those formed more than 50 Ma ago. This suggests that the mantle wedge beneath this segment of the Andes was not typical of the upper mantle. In Section 4.3.3, we described how the Rb/Sr ratio in the mantle wedge is a result of two processes; partial melting which decreases the ratio in the mantle, and influxed fluids which increase the ratio. Perhaps during the early life of an active continental margin, such as is illustrated by igneous rocks from northern Chile that formed more than 50 Ma ago, the effect of lowering the Rb/Sr ratio in the mantle from partial melt extraction outweighed the effects of fluids from the subducted slab.

ITQ 21

Is there any systematic difference between the initial Sr-isotope ratios of basaltic composition and those from more silica-rich compositions emplaced more than 50 Ma ago on Figure 4.36? What is the significance of your observation?

The strontium isotope evidence presented in Figure 4.36 therefore suggests that all magmatism prior to 50 Ma represented new additions to the crust in this segment of the Andes.

ITQ 22

Describe the change in initial Sr-isotope ratios for magmas younger than 50 Ma, and comment on its significance.

Geologists working on the tectonics of the Andes have established that the crust beneath the central Andes thickened about 50 Ma ago. As active continental margins mature, thickening of the crust often results from the emplacement of basaltic intrusions into the base of the crust (a process known as underplating). You may remember from Block 1 (Section 1.12.5) that because of the high concentrations of heat-producing elements (U, Th and K), some 40% of the surface heat flow is derived internally from radiogenic isotopes. Consequently, a thick continental crust will lead to a steep geothermal gradient. This in turn will lead to melting of the crust, particularly if the crust is also being heated by magmas derived from the mantle wedge. So the pattern of initial Sr-isotope ratios shown in Figure 4.36 can be related to the broad tectonic evolution of the central Andes. Magmas formed at the active continental margin thickened the crust from beneath, which in turn caused melting of continental rocks and hence produced melts with high initial $^{87}Sr/^{86}Sr$ ratios.

In summary, Figure 4.36 tells the isotope geochemist that crustal growth occurred between 200 and 50 Ma through the intrusion of mantle-derived magma. Crustal remelting is recorded in younger magmas, formed within a thickened crust.

In general, the isotopic evidence available from active continental margins suggests that such tectonic settings do provide sites for continental growth during the early period of their evolution. The addition rate of continental crust over the past few hundred million years is estimated to have averaged about $1 km^3$ per year at active continental margins worldwide. Crustal remelting becomes more important as the margin matures and the crust thickens. Overall, about 70% of magmas generated at active continental margins are believed to represent new additions to the crust from the mantle wedge and the rest, mainly granites, are a result of partial melting of crustal rocks.

SUMMARY OF SECTION 4.3

- Magmatism at active continental margins is characterized by more silica-rich compositions than at island arcs.

- Magmas of granitic compositions may be formed either by fractional crystallization of basic magmas or from partial melting of crustal rocks.

- The ternary system Qz–Ab–Or is characterized by a liquidus surface that slopes down towards a point generally referred to as the granite minimum. Consequently:

  Fractional crystallization of all samples in this system drives the composition of the residual liquid towards that of the granite minimum irrespective of their initial composition.

  Conversely, on heating a rock consisting largely of quartz and alkali feldspar, the first liquid to form is also close in composition to the granite minimum.

- The trace-element composition of granitic rocks can be modelled either by extreme fractional crystallization, or by a small degree of partial melting.

- Some isotopes are unstable and they decay, emitting radiation. The process of radioactive decay is described by a law which states that the rate at which the number of radioactive atoms decreases is proportional to the number of parent atoms present. The decay constant (λ) is *constant* for any particular isotope species.

- Isotope decay schemes, such as $^{87}Rb \rightarrow {}^{87}Sr$, may be used to determine the ages of both rocks and minerals. Ages are calculated using an isochron diagram ($^{87}Sr/^{86}Sr$ against $^{87}Rb/^{86}Sr$) on which samples of the same age and initial Sr-isotope ratio plot on a straight line whose slope is approximately equal to λt.

- The initial Sr-isotope ratio, $(^{87}Sr/^{86}Sr)_o$, of a sample is the isotope ratio at the time of its formation. It reflects both the age and the chemistry (as mirrored by different Rb/Sr ratios) of the source rock(s) from which the sample was derived. Consequently, initial Sr-isotope ratios are widely used to distinguish mantle-derived from crustal-derived melts.

OBJECTIVES FOR SECTION 4.3

When you have completed this Section, you should be able to:

4.1 Recognize and use definitions and applications of each of the terms printed in the text in bold.

4.7 Recognize and interpret the range of magmatic compositions formed at active continental margins.

4.8 Describe equilibrium and non-equilibrium crystallization and melting in the binary system Or–Ab.

4.9 Discuss partial melting and fractional crystallization in the synthetic granite system Qz–Ab–Or.

4.10 Infer the changes to the concentrations of trace elements in a magma during either fractional crystallization or partial melting, given the appropriate partition coefficients and graphical plots.

4.11 Describe the decay of ^{87}Rb to ^{87}Sr, and understand what happens to the different isotope ratios of Rb and Sr with the passage of time.

4.12 Plot $^{87}Sr/^{86}Sr$ and $^{87}Rb/^{86}Sr$ results on an isochron diagram, and determine the age of the samples and their initial Sr-isotope ratio.

4.13 Understand the assumptions that underlie the interpretation of the calculated age of a suite of rocks that define an isochron.

4.14 Use initial Sr-isotope ratios to evaluate the likely source of particular igneous rocks.

4.15 Interpret and discuss the evidence for crustal growth at active continental margins given appropriate geochemical data.

Apart from Objective 4.1, to which they all relate, the eleven ITQs in this Section test the Objectives as follows: ITQ 12, Objective 4.7, ITQ 13, Objective 4.8, ITQ 14, Objective 4.9, ITQs 15 and 16, Objective 4.10, ITQ 17, Objective 4.11, ITQs 18 and 19, Objective 4.12, ITQ 20, Objective 4.14, ITQs 21 and 22, Objective 4.15.

You should now do the following SAQs, which test other aspects of the Objectives.

SAQS FOR SECTION 4.3

SAQ 6 (*Objective 4.7*)

Give two objections based on Figure 4.22 to the suggestion that most high-SiO_2 igneous rocks generated along a destructive plate margin such as the Andes were produced by fractional crystallization of basaltic liquids alone. Give your reasons in a couple of sentences.

SAQ 7 (*Objective 4.9*)

A sample consists of 20% quartz, 70% albite and 10% orthoclase. Plot this on the triangular diagram SiO_2–$NaAlSi_3O_8$–$KAlSi_3O_8$ (Figure 4.25) and answer the following questions:

(a) On cooling from a temperature of 860 °C, which mineral crystallizes first?

(b) Do those first crystals contain more or less orthoclase than the coexisting liquid?

(c) At what temperature do the first crystals appear?

(d) Assuming that complete equilibrium is maintained throughout crystallization, does the feldspar become richer in albite or orthoclase as the temperature falls?

(e) At approximately what temperature will the first crystals of quartz appear?

(f) If crystallization takes place under equilibrium conditions, will the last drop of liquid have the same composition as the granite minimum (M)?

SAQ 8 (*Objective 4.12*)

Plot the results for the six rocks in Table 4.8 in Figure 4.37. Draw a straight line through at least five points: calculate the slope of the isochron and hence the age of the samples. (λ for $^{87}Rb = 1.42 \times 10^{-11} a^{-1}$.)

Table 4.8 Data for SAQs 8–10.

Sample	$^{87}Sr/^{86}Sr$	$^{87}Rb/^{86}Sr$
1	0.7130 ± 0.0002	0.12 ± 0.01
2	0.7187 ± 0.0002	0.95 ± 0.01
3	0.7245 ± 0.0002	1.75 ± 0.01
4	0.7262 ± 0.0002	2.00 ± 0.01
5	0.7250 ± 0.0002	2.50 ± 0.01
6	0.7310 ± 0.0002	2.70 ± 0.01

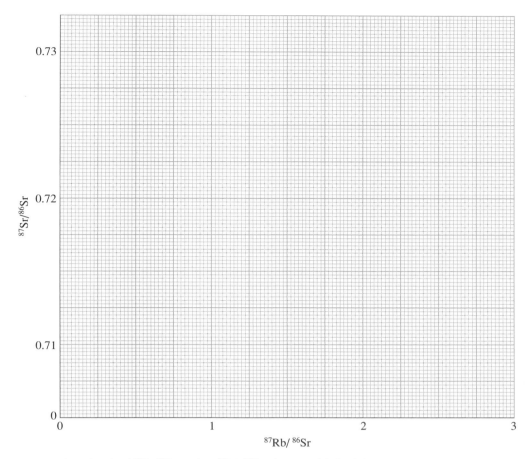

Figure 4.37 Graph of $^{87}Sr/^{86}Sr$ against $^{87}Rb/^{86}Sr$ (for use with SAQ 8).

SAQ 9 (*Objective 4.15*)

Can you suggest *two* reasons why one of the samples in Table 4.8 does not fall on the isochron with the other five samples?

SAQ 10 (*Objective 4.14*)

What is the initial Sr-isotope ratio of the five samples you have plotted on the isochron in Figure 4.37? If they were samples of granite from Greenland, would their initial Sr-isotope ratio indicate that they were derived from (a) the upper mantle, (b) the metamorphic granites of Figure 4.31, (c) another crustal source?

SAQ 11 (*Objective 4.15*)

A group of granites from Arabia were dated at 500 ± 10 Ma with initial Sr-isotope ratios that varied from 0.703 8 to 0.704 2. Does this provide evidence for, or against, growth of continental crust beneath Arabia (a) 500 Ma ago, (b) today?

4.4 COLLISION ZONES

In the Course so far, we have examined the origins of magmas that have formed at three types of plate margins: mid-ocean ridges, island arcs and active continental margins. You may have noticed that these processes have been introduced in a logical sequence, each one involving more silica-rich magmas than the last. However, we have not yet exhausted all possible environments in which magmas contribute to the evolution of continental crust.

The evolved magmas that are generated at island arcs and active continental margins form an irreversible addition to the crust. Continents will grow through geological time, and the motions of plate tectonics will cause the expanding continents to be shunted around like scum on the surface of a pond. It is inevitable that collisions will occur at some stage between two continental masses. Indeed, if the subducted plate at an active continental margin incorporates a passive continental margin, then two continents must converge on a collision course (Figure 4.38).

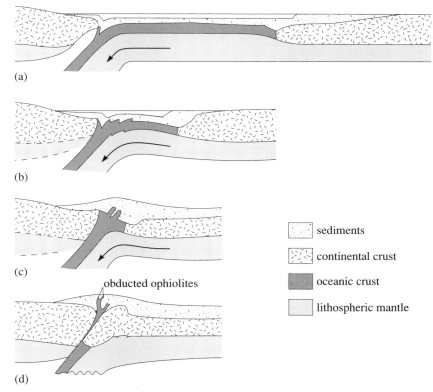

(a)

(b)

(c)

obducted ophiolites

(d)

sediments

continental crust

oceanic crust

lithospheric mantle

Figure 4.38 Sequence of events leading to continental collision: (a) subduction causing contraction of ocean basin; (b) pre-collision with deformation of continental margin sediments; (c) early collision with deformation of oceanic lithosphere; (d) late collision and the beginning of crustal thickening.

The consequences of a collision between two plates of continental lithosphere are far-reaching. Neither continent can be subducted far due to the buoyancy of continental crust, and so the forces that drive the plate movement prior to collision are brought to bear directly on the continental lithosphere itself. Under such strong compressive forces, the lithosphere contracts through folding, faulting and deforming the rocks that make up the continental crust. In response to these processes, the crust will also thicken. Because of the buoyancy of continental crust, regions of thickened crust will be uplifted. That is one reason why some of the high mountain ranges of the Earth are underlain by crust of great thickness. The crust beneath the Himalayas and the Tibetan Plateau (about 55–70 km) is approximately double the normal thickness of average continental crust. Other mountain ranges that mark sites of continental collision include the Alps and the Urals. However, it should be emphasized here that collision is only one way of thickening the crust, and the thick crustal root that underlies some of the Andes (Figure 4.33) is a result of magmas derived from the mantle wedge and intruded into the base of the crust.

As the crust thickens, the rocks in the lower part of the crust will be subjected to increasing pressures. At the same time, the geothermal gradient will steepen due to the effects of thick sequences of continental rocks with high concentrations of the heat-producing elements (U, Th and K). These changes in pressure and temperature will result in mineral recrystallization (metamorphism) beneath collision zones. If the temperature rises above the solidus for common sedimentary rocks, then the crust will start to melt and granitic magmas are formed. But unlike the granites found at active continental margins, collision-zone granites are not associated with less-evolved plutonic rocks like gabbros, diorites or even granodiorites. Partial melting in a collision zone requires no contribution from the mantle either directly as basaltic magmas or even indirectly as a heat source.

The final Section in this Block will examine metamorphism and metamorphic processes, culminating in melting of crustal rocks, looking particularly at crustal melts from the Himalayan collision zone.

4.4.1   METAMORPHIC PROCESSES

Metamorphism is a term used to describe the changes that affect existing rocks when they are subject to a change in pressure and/or temperature. It usually refers to chemical and physical reactions that take place in the *solid state*, though many metamorphic reactions are greatly assisted by the presence of any fluid that may be present along the grain boundaries or released by the reaction itself. Metamorphism covers a great range of processes, from subtle changes in the structure of clay minerals that occur in sediments after their burial, to the mineral reactions that occur deep in the crust at high pressures and temperatures. The field of metamorphism extends to the partial melting of rocks, so that as you will see, there is overlap between the physical conditions under which igneous and metamorphic processes occur.

The simplest form of metamorphic reaction involves the change of a single mineral from one structure to another. If two or more minerals have the same composition, but contrasting structures, they are known as polymorphs. You are familiar with the polymorphs of carbon, diamond and graphite, from Block 3 (Section 3.3.3), and their **stability fields** are plotted in Figure 4.39. Both polymorphs may coexist along the phase boundary that separates their stability fields.

Increasing the pressure of a rock will lead to reactions by which minerals of low density will be replaced by minerals of comparatively high density. Diamond is denser than graphite ($3\,500\,\text{kg m}^{-3}$ compared with $2\,000\,\text{kg m}^{-3}$), and so high pressures favour the graphite $\rightarrow$ diamond reaction. In other words, the transition from graphite to diamond results in a decrease in volume.

The graphite $\rightarrow$ diamond reaction is an example of a **geobarometer**, because the occurrence of either phase provides a constraint on the pressure at which a rock has formed. For example, if we know that graphite-bearing rocks are formed at a temperature of $1\,000\,°C$, then the stability field of graphite (Figure 4.39) tells us that the maximum pressure the rock formed at was about $4\,\text{GN m}^{-2}$. It is not a very useful barometer though, because graphite is stable for all crustal pressures. Much more useful for petrologists who study crustal rocks are minerals that undergo metamorphic reactions under conditions commonly found within the crust. A particularly useful example is provided by three minerals that occur as minor components in many rocks that have aluminium-rich compositions: the **aluminosilicates**.

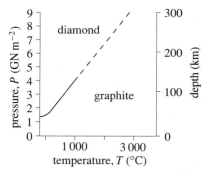

Figure 4.39   Stability fields of graphite and diamond in pressure–temperature space.

All three minerals are polymorphs with the same chemical formula, Al_2SiO_5, but their internal structures and external appearances are markedly different. The three polymorphs are called **kyanite**, **andalusite** and **sillimanite**, and these are illustrated in Plates 4.3, 4.4 and 4.5. Each aluminosilicate is typical of a different range of pressure–temperature conditions and their stability fields are shown in Figure 4.40. Such diagrams are the results of experiments performed in the laboratory, and the precise position of each boundary is subject to experimental uncertainty. This is particularly true for polymorphic reactions that have sluggish reaction rates.

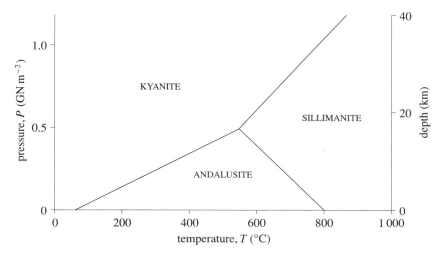

Figure 4.40   The stability fields of the three aluminosilicate minerals in pressure–temperature space.

ITQ 23

From Figure 4.40, estimate the relative densities of the three aluminosilicate minerals.

The stability fields of the aluminosilicates (Figure 4.40) can be used to provide information of the physical conditions under which metamorphic rocks formed. For example, we know that the sillimanite-rich rock shown in Plate 4.4 could not have formed at temperatures less than about 580 °C (the minimum temperature for the sillimanite stability field). We also know that the andalusite in Plate 4.5 could not have formed at pressures greater than about $0.5\,GN\,m^{-2}$ (Figure 4.40). Since these andalusites are from baked sediments at the contact of a large granite in Cumbria, we now know that the granite was intruded at pressures less than $0.5\,GN\,m^{-2}$, equivalent to depths less than about 15 km. Thus, the aluminosilicates allow us to divide up the pressure–temperature space of metamorphic rocks into high-pressure (kyanite), low-pressure (andalusite) and high-temperature (sillimanite) regimes. Other minerals can be used in similar ways, and the conditions under which metamorphic rocks formed can sometimes be pin-pointed quite accurately.

So far, we have concentrated on the consequences of increasing pressure on mineral stability. We have also considered only single minerals and their polymorphs. In most natural rocks, several minerals take part in a metamorphic reaction in response to changing pressure *and* temperature. For example, let us consider the minerals calcite ($CaCO_3$) and quartz (SiO_2). If a limestone containing mostly calcite with minor quartz is heated, a useful metamorphic reaction will occur:

$$CaCO_3 + SiO_2 = CaSiO_3 + CO_2 \qquad \text{(Equation 4.13)}$$

calcite   quartz   wollastonite

In the metamorphosed limestone (known as a marble) small needles of the mineral wollastonite ($CaSiO_3$) are found along the grain boundaries of calcite and quartz (Figure 4.41).

The stability fields for calcite + quartz and for wollastonite + CO_2 are shown in Figure 4.42. Notice that the wollastonite + carbon dioxide assemblage is favoured by high temperatures.

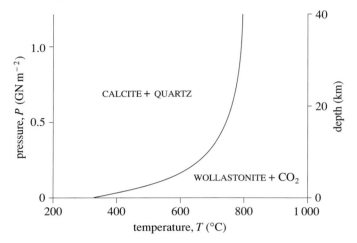

Figure 4.42   Boundary curve for the reaction calcite + quartz = wollastonite + CO_2.

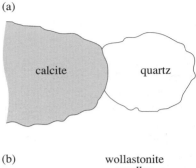

(a)

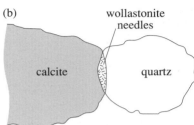

(b)

Figure 4.41   Sketch to show the formation of wollastonite ($CaSiO_3$) needles during metamorphism of limestone. The two diagrams (a) and (b) show the minerals present before and after metamorphism respectively.

Just as high pressures stabilize minerals with high densities, so high temperatures stabilize minerals and gases with a high degree of disorder in their molecular structure. For polymorphs such as the aluminosilicates, the mineral with the more disordered state will always lie on the high temperature side of a phase boundary. Thus, sillimanite has the most disordered molecular structure of the three Al_2SiO_5 polymorphs, and kyanite has the lowest (Figure 4.40). Since the structures of gases are considerably more disordered than those of solids, any mineral reaction that releases a gas, such as given by Equation 4.13, will be triggered by an increase in temperature. In other words, the phase boundary for the reaction will be steep (at least at higher pressures) as we can see from Figure 4.42. The steep phase boundaries that describe reactions that produce CO_2 (decarbonation reactions), or H_2O (dehydration reactions) provided excellent **geothermometers**.

ITQ 24

A limestone contains wollastonite, calcite and quartz. The adjacent rock is an aluminous sediment that contains kyanite. By considering both Figures 4.40 and 4.42, estimate the minimum pressure at which these rocks have formed.

So far, we have discussed how different pressures and temperatures can change the minerals that are stable in a rock. It is also true that *different rock compositions will allow different minerals to form under the same pressure and temperature conditions*. When we were describing rocks that contained different aluminosilicates, the chemical composition of the rocks must have been rich in aluminium to form any aluminosilicates at all. Some compositions like sandstone, made up of just quartz, are quite sterile in terms of mineral reactions because quartz is stable for all normal conditions encountered in the crust. In contrast, aluminium-rich sediments, known as **mudstones**, provide the raw materials for many mineral reactions. Mudstones are made up of clay minerals which are prone to great changes during metamorphism on account of their chemical complexity. We shall now look at the metamorphism of a mudstone under increasing pressure and temperature.

First, we consider the textural changes that are summarized in Table 4.9. The fine-grained clay minerals that make up a mudstone are deposited flat on the floor of the stream or lake. On compaction, the flakes are rotated to align at right-angles to the compacting load, giving the rock a layered appearance. When this compaction is accompanied by heating, the clay minerals begin to recrystallize, and the process is promoted by hot water-rich fluids being forced through the sediments caused by the compacting load of overlying sediments. In this way, new micaceous (mica-rich) minerals form. The micas will be either biotite, if iron-rich, or muscovite, if aluminium-rich. All micas are sheet-like or platey in shape (Appendix 1, Block 3) and grow perpendicular to the direction of compression acting on the rock in which they grow. This stress is usually vertical due to the compacting load, but may vary due to non-vertical stress induced by plate movements.

Slates are the result of a small increase in pressure and temperature acting on a mudstone. The minerals are still very fine grained but they are largely micaceous: the preferred orientation of the micas within the rock gives it a well-defined cleavage (exploited in the manufacture of roofing slates, for example). Further metamorphism of mudstones results in an increased grain size. Coarsely crystalline rocks with an aligned micaceous texture are known as schists (Sample 9 in the Kit). But micas are not the only minerals that form during metamorphism.

Table 4.9 Classification of metamorphosed sediments

Rock type	Appearance	Grain size
Migmatite	Separation of granitic lenses within darker layers (usually biotite-rich) or amphibole	Coarse
Gneiss	Widely spaced layers with alternations between mica, amphibole or pyroxene-rich layers and quartz, plagioclase-rich layers	Coarse
Schist	Moderately spaced, sub-parallel planes, characteristically with abundant mica	Medium
Slate	Very closely spaced, almost perfectly flat planes	Fine

One of the aluminosilicate minerals can also result from a reaction involving micas. For example, a common dehydration reaction in schists is:

$$\underset{\text{muscovite}}{KAl_3Si_3O_{10}(OH)_2} + \underset{\text{quartz}}{SiO_2} = \underset{\text{alkali feldspar}}{KAlSi_3O_8} + \underset{\text{aluminosilicate}}{Al_2SiO_5} + H_2O \quad \text{(Eqn 4.14)}$$

Higher temperatures and pressures will result in a segregation between quartz + feldspar and ferromagnesian minerals, such as biotite or amphibole. Since the former are light in colour and the latter are dark, this gives the rock a banded appearance. Such rock is known as a **gneiss** (which is pronounced 'nice'). Not only mudstones, but other rock-types such as granites and gabbros can form gneisses (Plate 4.6) when metamorphosed to sufficiently high temperatures and pressures. The dark layers may include micas, amphiboles, or pyroxenes depending on the degree of metamorphism and on the bulk composition of the rock.

It is under conditions close to the breakdown of muscovite that a gneiss may begin to melt. This is illustrated in Figure 4.43 (*overleaf*), where both the granite melting curve and the reaction boundary defined by Equation 4.14 are plotted. The granite melting curve shown is for melting in the presence of quartz, plagioclase, alkali feldspar and water.

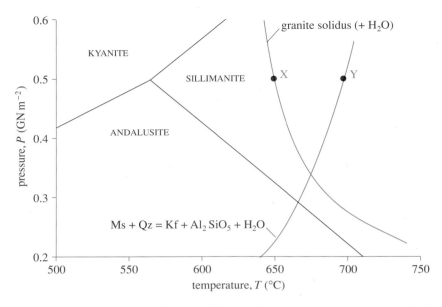

Figure 4.43 Solidus for melting of granite (in the presence of water) and the muscovite and quartz dehydration reaction (Equation 4.14). Also shown are the stability fields for the aluminosilicates. Ms, muscovite; Qz, quartz; Kf, alkali feldspar.

■ If we follow the effects of increasing temperature on a typical metamorphosed mudstone at pressures of $0.5\,GN\,m^{-2}$, what will happen when point X is reached at about 650 °C (Figure 4.43)?

❑ X lies on the water-present granite solidus, but no reaction will occur. In order to form a granite magma at this point, the three minerals quartz, plagioclase, and alkali feldspar together with water must be present in the rock. Metamorphic rocks at temperatures of several hundred degrees Celsius rarely contain a free water phase and the melting of granite in the absence of water occurs at much higher temperatures. Also absent from many metamorphosed sediments is the mineral alkali feldspar.

■ What will happen when point Y is reached?

❑ At Y, the muscovite dehydration reaction is crossed. This will generate some water (see Equation 4.14) and also alkali feldspar. Since quartz and plagioclase are common minerals in sediments, all four phases (plagioclase, alkali feldspar, quartz and water) required for forming a granite minimum-melt composition are now present. Some partial melting will therefore occur.

■ Equation 4.14 also has Al_2SiO_5 as a product. Which aluminosilicate will be stable?

❑ From Figure 4.43, Y lies within the sillimanite field, so sillimanite will form.

If only a small proportion of melt is generated, the granite magma will remain in the rock but be segregated from the micaceous layers. When these melts crystallize, they will form lenses of quartz, plagioclase and alkali feldspar. Typically such lenses are a few centimetres thick. These mixed rocks are called **migmatites** (Plate 4.7).

More extensive melting will allow the melts to coalesce and migrate upwards as a granite pluton. With high enough temperatures, the hydrous phases in all crustal compositions will be dehydrated. Muscovite will melt, leaving an aluminosilicate-rich residue. Biotite will melt to leave a garnet-rich or pyroxene-rich residue. If igneous rocks of basic and/or intermediate composition are metamorphosed, amphibole will break down to leave a pyroxene-rich residue. The residual dehydrated rocks are known as **granulites**, and they characterize much of the lower crust where conditions are too hot for micas or amphiboles to be stable.

We have now returned to the realm of igneous petrology. But before we leave this introduction to metamorphism, we should emphasize that metamorphism can occur in any tectonic setting, not just collision zones, in response to changes in temperature or pressure. Andalusite, the low-pressure aluminosilicate, is commonly formed where aluminous sediments are intruded by magma at shallow depths irrespective of the tectonic setting in which the magma forms (for example, the andalusite shown in Plate 4.5 is from a metamorphosed sediment caused by the intrusion of a granite formed at an active continental margin). Indeed, the continental crust can thicken beneath active continental margins, such as the Central Andes, causing metamorphism and deformation very similar to that found in collision zones although not over such a regional extent.

Metamorphism also occurs within the oceanic crust as basaltic compositions are subjected to higher temperatures at the deeper levels. And from Section 4.2.1, we know that eclogite is the result of the high temperatures and pressures that subducted oceanic lithosphere can experience. But it is in regions of thickened continental lithosphere that crustal rocks are subjected to unusually high temperatures and pressures over a wide area. It is here that we are most likely to find evidence for crustal melting.

4.4.2 METAMORPHISM AND MAGMATISM IN THE HIMALAYAS

The Himalayas today form a spectacular mountain range that defines the northern limit of the Indian plains and the southern limit of the Tibetan plateau. On the Smithsonian Map, you will see that an arc of earthquake epicentres coincides with the Himalayan range, suggestive of a rather diffuse plate boundary. From Block 2 (Section 2.5.4), you may recall that the Himalayas mark the site of collision between two continents which continue to converge. The initial collision between two great continental plates, India and Eurasia, occurred between 40 and 55 Ma ago. Continued northwards movement of India led to considerable thickening of continental crust and uplift of both the Himalayas and the Tibetan Plateau. All that is left of the once great ocean that separated India and Eurasia is a few narrow slivers of ophiolite (Figure 4.44) that define the suture between the these two plates. One way in which India continues its northwards progression is by movement along northward dipping thrusts resulting in the earthquake epicentres marked on the Smithsonian Map.

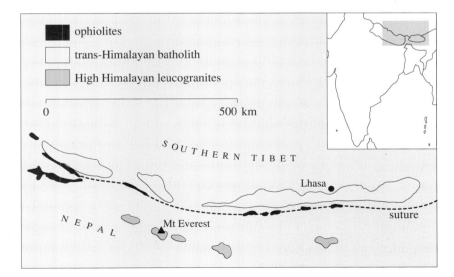

Figure 4.44 Geological sketch-map across the Himalayas and southern Tibet showing the distribution of the trans-Himalayan batholith and the High Himalayan granites.

The Himalayas provide evidence of two contrasting examples of magmatism that can be found in collision zones. The first of these is the trans-Himalayan batholith which is emplaced into the southern margin of the Asian plate, just north of the suture that marks the boundary between pre-collision Asia and India (Figure 4.44). This forms a vast igneous arc, over 3 000 km long, and 50 km wide (only a segment of it is shown in Figure 4.44), made up of plutons of granites, granodiorites, diorites and a small proportion of gabbros. Rb–Sr isochron ages from the plutons cover a range from about 110 Ma to 40 Ma. These magmas intrude volcanic rocks of similar age and also sediments that were deposited along a continental margin between 100 and 200 Ma ago. The initial Sr-isotope ratios of several plutons from the trans-Himalayan batholith are plotted in Figure 4.45.

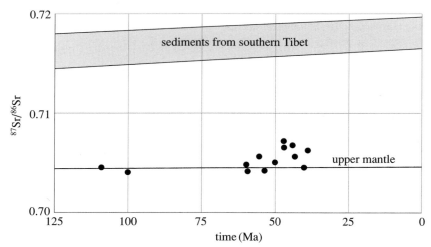

Figure 4.45   Sr-evolution diagram of plutons from the trans-Himalayan batholith and of the sediments that they intrude.

ITQ 25

From Figure 4.45, test the following hypotheses.

(a) The trans-Himalayan batholith results from melting of the sediments it intrudes.

(b) The trans-Himalayan batholith results from fractional crystallization of mantle melts.

(c) The trans-Himalayan batholith results from contamination of mantle melts by crustal rocks.

You may ignore possible isotopic heterogeneities in the upper mantle.

We can now draw together tectonic and geochemical evidence for the formation of the trans-Himalayan batholith. The magmas were emplaced during subduction of oceanic lithosphere, but magmatism ceased after collision between India and Asia 40–50 Ma ago. They were therefore emplaced above an active subduction zone at an active continental margin (Figure 4.38(a)). From the range of rock-types (granites to gabbros), the suite of magmas appears to be similar to magmas generated in the Andes, and detailed geochemical studies support such an analogy. Therefore, the trans-Himalayan batholith is a pre-collision example of magmatism at an active continental margin. Once collision occurred, subduction of oceanic lithosphere stopped, cutting off the supply of fluids into the mantle wedge. As a result, no further melting in the mantle was possible.

The second example of Himalayan magmatism contrasts strongly with the trans-Himalayan batholith. A series of light-coloured granites (the High Himalayan leucogranites) are emplaced as lenses and sheets into the sediments of the Himalayan mountains south of the suture (Figure 4.44). They are known as leucogranites because they are light in colour (Plate 4.8), being composed almost entirely of quartz, plagioclase and alkali feldspar. All the intrusions are granitic in composition containing 70–75% silica. Rb–Sr isotope data from several of these granites indicate an age of 20 ± 5 Ma and an initial Sr-isotope ratio varying between 0.730 and 0.760. The granites are therefore intruded well after collision and have a much higher initial Sr-isotope ratio compared to any analysed samples from the trans-Himalayan batholith (Figure 4.45).

ITQ 26

Suggest a possible source for the High Himalayan leucogranites (based on the isotope data and rock-type described so far) and relate this to what you know of Himalayan tectonics.

The High Himalayan leucogranites are intruded, not into unmetamorphosed sediments like the trans-Himalayan batholith, but into migmatites formed from the partial melting of much older aluminous sediments (Plate 4.7). The sediments found to the north of the suture in southern Tibet have $(^{87}Sr/^{86}Sr)_o < 0.720$ (Figure 4.45), but further south in the Himalayas aluminous sediments provide extremely high $(^{87}Sr/^{86}Sr)_o$ of 0.730–0.780. Careful examination of the minerals in the dark parts of the migmatites shows that muscovite is breaking down to form sillimanite by reaction with quartz (Equation 4.14). This reaction provides alkali feldspar and a hydrous fluid which, when combined with quartz and plagioclase from the sediment, will form a granite melt.

The granite melting reaction (in the presence of water) can be described by

$$SiO_2 + NaAlSi_3O_8 + KAlSi_3O_8 + H_2O = \text{granite melt} \qquad \text{(Eqn 4.15)}$$
$$\text{quartz} \quad \text{albite} \qquad \text{alkali feldspar}$$

Combining Equations 4.14 and 4.15, we obtain

$$\text{muscovite} + \text{quartz} + \text{albite} = \text{sillimanite} + \text{granite magma}$$
$$\text{(Equation 4.16)}$$

Therefore the mineral reaction that is observed in the migmatite provides a possible source for the granite.

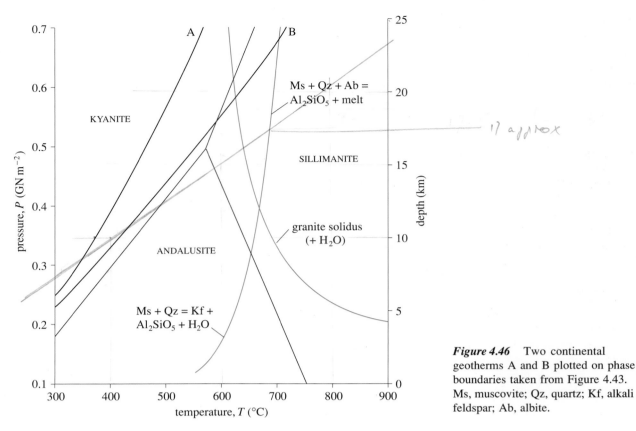

17 approx

Figure 4.46   Two continental geotherms A and B plotted on phase boundaries taken from Figure 4.43. Ms, muscovite; Qz, quartz; Kf, alkali feldspar; Ab, albite.

Why rocks should melt at all in collision zones can be understood from Figure 4.46. On it, the aluminosilicate stability fields, the muscovite + quartz breakdown reaction and the granite melting curve have been transferred from Figure 4.43, and superimposed on the diagram are two geotherms, A and B. Geotherm A is a typical geotherm through continental crust. Geotherm B is an elevated geotherm that is a consequence of internal heating within thickened continental crust.

ITQ 27

(a) At which depths would you expect melting to occur in response to (i) geotherm A and (ii) geotherm B of Figure 4.46? (b) Which polymorph of Al_2SiO_5 will be formed by the melting reaction?

Geologists who have studied one of the High Himalayan granites observed that although the granite was emplaced into migmatites, a second set of aluminous metasediments (mica schists) was also present at deeper structural levels (Figure 4.47). Because they come from deeper in the crust, these sediments contained kyanite, not sillimanite. A debate arose between those who maintained the granite resulted from melting migmatites *in situ* and those who suggested it had melted out of sediments deeper in the crust and the magma had subsequently risen and been intruded into the migmatites. This may seem a very academic discussion, but in fact it proved to be important, as we shall explain.

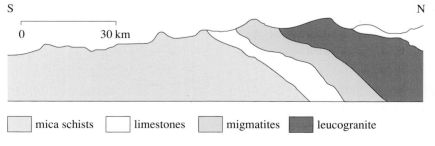

Figure 4.47   Vertical north–south section through a High Himalayan leucogranite intruded into metamorphosed sediments.

It is widely believed that many granitic magmas rise through the crust by virtue of their low density (i.e. through buoyancy). Extraction of a melt in this way will happen more readily when a high degree of melting occurs. Indeed, for granite magma, a melt fraction less than 0.3 is unlikely to be able to leave its source by this means.

ITQ 28

The Himalayan leucogranites have an average Rb content of 300 ppm. If the bulk partition coefficient (D) for Rb in a mica-rich sediment is 0.44, what melt fraction (F) will be required to produce the melt from a sediment with Rb contents of 150 ppm?

A low melt fraction of little over 0.1 is confirmed from a range of geochemical tests for the High Himalayan granites. This means that such melts could not have left their source through buoyancy. They would remain in place as sheets and lenses in migmatites.

ITQ 29

A seven-point Rb/Sr isochron diagram for a Himalayan granite is plotted on Figure 4.48. The Sr-isotope evolution lines for the mica schist and migmatite of Figure 4.47 are plotted on Figure 4.49.

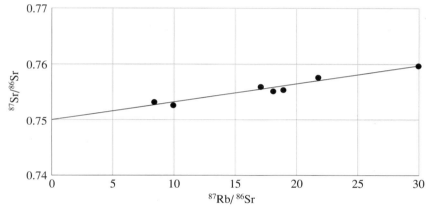

Figure 4.48 For use with ITQ 29. Isochron diagram for a suite of seven samples from a High Himalayan leucogranite.

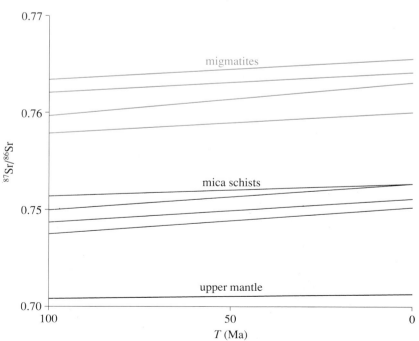

Figure 4.49 For use with ITQ 29. Sr-isotope evolution diagram for mica schists and migmatites from the Himalayas.

Determine (a) the age of the granite, and (b) whether the mica schist or the migmatite is the likely source for the granite.

The answer to ITQ 29 is interesting because it implies that the granite magma has been able to leave its source and was emplaced into the migmatites about 10 km higher up the sedimentary sequence.

■   How did such a low melt fraction (remember $F = 0.1$) leave its source?

❑   The answer is that the melt could not have migrated in response to buoyancy alone. One explanation for extracting low melt fractions is that the region was under extension at the time of melt formation. Extensional stress in the crust will create cracks and fractures that allow melts to migrate upwards more easily through dykes.

Of course, the dominant stress field in a collision zone is one of compression. But within this regional picture, extension can, and does, occur, and faults formed during extension (normal faults) are one of the main mechanisms for uplifting the mountains themselves. Indeed, after geochemists suggested that melts may have been emplaced under extension in the Himalayas, major extensional faults have been identified and are now known to have been active at the time of granite emplacement. If you want to know more about the tectonics of collision zones, you should study the third level Earth science course 'Understanding the Continents'.

SUMMARY OF SECTION 4.4

*   Collision between two continental plates results in thickening and metamorphism of the crust.

*   Varying pressures and temperatures in the Earth's crust results in reactions between metamorphic minerals.

*   High pressures favour high-density minerals, and therefore reactions involving large changes in densities provide geobarometers. Geothermometers are provided by reactions that result in dehydration or decarbonation.

*   The stability field of a mineral or group of minerals may be represented on a pressure–temperature diagram which then provides information on the pressure or temperature to which a rock containing the mineral(s) has been subjected.

*   Increasing temperatures and pressures of sediments that contain clay minerals result in the formation of slates, schists, gneisses and migmatites.

*   The reaction of quartz and muscovite can produce granite melts in sediments of aluminous compositions.

*   In the Himalayas, pre-collision magmatism is represented by the trans-Himalayan batholith that was intruded between 110 and 40 Ma. It is typical both in rock-type and geochemistry of magmatism formed at an active continental margin.

*   The High Himalayan leucogranites (intruded at about 20 Ma) result from partial melting of mica schists in response to thickening of the continental crust through collision.

OBJECTIVES FOR SECTION 4.4

When you have completed this Section, you should be able to:

4.1 Recognize and use definitions and applications of each of the terms printed in the text in bold.

4.16 Understand the significance of density and molecular disorder in determining the stability field of a mineral or group of minerals and hence identify useful geothermometers and geobarometers.

4.17 Interpret pressure–temperature diagrams that illustrate the stability fields of polymorphs or mineral assemblages.

4.18 Identify the contribution of crustal melting to granite formation in collision zones from Sr-isotope studies.

4.19 Integrate tectonic and geochemical studies in collision zones to explain metamorphism and its relationship with magmatism.

Apart from Objective 4.1 to which they all relate, the seven ITQs in this Section test the Objectives as follows: ITQ 23, Objective 4.16, ITQ 24, Objective 4.17, ITQ 25, Objective 4.18, ITQs 26–29, Objective 4.19.

You should now do the following SAQs, which test other aspects of the Objectives.

SAQS FOR SECTION 4.4

SAQ 12 (*Objectives 4.16 and 4.17*)

Aragonite and calcite are both polymorphs of calcium carbonate ($CaCO_3$) and their stability fields are given in Figure 4.50. (a) Which of the two polymorphs is likely to be denser? (b) Which of the two polymorphs is likely to have the more disordered molecular structure?

SAQ 13 (*Objective 4.18*)

Using the Himalayas as an example, does the collision between two continents result in increased or decreased crustal growth rates?

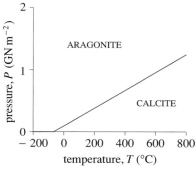

Figure 4.50   The stability fields of calcite and aragonite (for use with SAQ 12).

4.5 GRANITIC MAGMAS AND PLATE TECTONICS: CONCLUDING REMARKS

Fractional crystallization and partial melting are two processes that allow the geochemist to understand how the wide range of igneous rocks that make up the oceanic and continental crusts can be generated from a peridotite source. In this Block, we have illustrated how acid and intermediate magmas can form directly from mantle-derived melts leading to crustal growth, particularly at island arc and active continental margins. Acid magmas form in regions of thickened crust such as at collision zones, and also to a lesser extent at active continental margins, by partial melting of sediments, thus exemplifying crustal reworking. But it would be a mistake to leave the impression that acid magmas *only* form in these environments.

Rhyolites, for example, occur in Iceland, which lies on a mid-ocean ridge. Although forming a tiny proportion of the total volume of magma generated in such a setting, it is possible to produce acid magmas wherever basic and intermediate magmas are trapped in magma chambers that allow a high degree of fractional crystallization to occur. Under unusual circumstances this may happen, not only at mid-ocean ridges but also at ocean islands as in Ascension in the Atlantic Ocean.

Moreover, crustal melting does not only occur within thick continental crust. It will happen whenever the temperature of crustal rocks exceeds that of the solidus for those rocks. Locally, melting can be seen in sediments around the margins of many basic intrusions. And perhaps more importantly, melting can occur wherever the lithosphere is stretched and thinned, thus bringing the asthenosphere closer to the surface and producing very steep geothermal gradients. Examples of granites formed in extensional environments include granites in southern Arabia, associated with the rifting of the Red Sea, and granites formed in the western isles of Scotland, associated with the opening of the Atlantic (as illustrated in video band VB 06 'Magmas in Scotland'). But if you bear these exceptions in mind, this Block has shown how the great majority of acid magmas are formed — at convergent plate boundaries.

ITQ ANSWERS AND COMMENTS

ITQ 1

(a)   At 100 km, wet melting of basalt can occur for geotherm 2 and probably even for the cold-slab model too (geotherm 1).

(b)   At 200 km, wet melting would occur for either geotherm but no water is available at such depths (Figure 4.3). However, dry melting of basalt would be possible with geotherm 2 (at such depths, basaltic compositions would have been metamorphosed to eclogite).

ITQ 2

Neither geotherm in Figure 4.4 crosses the dry peridotite solidus. However, wet melting only requires temperatures around 1 000 °C at depths of about 100 km. Such temperatures are reached in the mantle wedge, and water expelled from the subducted plate would therefore allow melting at these depths (Figure 4.3).

ITQ 3

Partial melting of mantle peridotite will result in a basaltic magma. Partial melting of the upper slab of basaltic composition will result in a magma that has a lower melting point than the parent basalt (i.e. andesite).

ITQ 4

The answers refer to Figure 4.51.

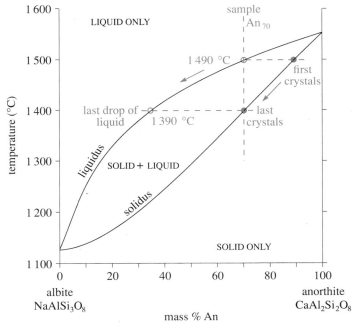

Figure 4.51   Answer to ITQ 4.

(a)   A sample with a composition An_{70} contains 70% anorthite and 30% albite, so the answer is 30% albite.

(b)   The first crystals appear when the total sample first plots on the liquidus. For a sample of An_{70}, this occurs at a temperature of 1 490 °C.

(c)   The first crystals plot on the *solidus* at that same temperature of 1 490 °C; their composition is therefore An_{89}.

(d)   The composition of the *total* sample does not change under equilibrium conditions; therefore it is still An_{70}.

(e)   The sample becomes completely solid when the total sample moves from the solid-and-liquid field to the solid-only field, that is, when it crosses the solidus. For a sample of An_{70}, this occurs at a temperature of 1 390 °C.

(f)   The last drop of liquid before crystallization is complete plots on the liquidus at that temperature of 1 390 °C. The composition of the liquid at 1 390 °C is An_{33}.

(g)   The composition of the crystals in equilibrium with the last drop of liquid is that of the total sample An_{70}.

ITQ 5

The answers are illustrated in Figure 4.52.

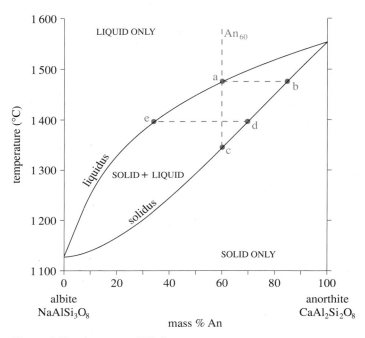

Figure 4.52   Answer to ITQ 5.

(a)   On cooling, a sample of An_{60} first intersects the liquidus at point a, and thus the first crystals plot on the solidus at the same temperature (point b). Their composition is An_{85}. The sample An_{60} leaves the solid-and-liquid field at point c, and under equilibrium conditions the composition of the last crystals is the same as the total sample, that is An_{60}.

(b)   If at 1 400 °C, the crystals (An_{75} at d) and the liquid (An_{35} at e) become isolated from each other, then we may consider their crystallization separately. If the continuing crystallization of the liquid is under equilibrium conditions, then the composition of the final crystals is the same as that of that liquid (An_{35}). Thus, when solidification of the *total* sample An_{60} is complete, there are crystals of An_{35} and An_{75}.

ITQ 6

The answer is illustrated in Figure 4.53.

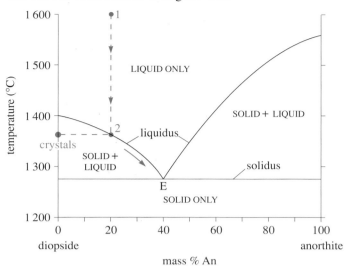

Figure 4.53 Answer to ITQ 6.

At 1 600 °C, the sample plots in the liquid-only field (point 1), and it remains there until the temperature falls to 1 360 °C, whereupon it plots on the liquidus (point 2) and the first crystals appear. They are all of the mineral diopside. As the temperature falls below 1 360 °C, the total sample moves through the solid-and-liquid field; the crystals plot on the solidus and the liquid moves down along the liquidus. More and more crystals of diopside are formed and thus the liquid becomes richer in anorthite, until at 1 275 °C crystals of anorthite start to appear, crystallizing along with diopside. The temperature remains constant at this point (the eutectic point, E) until the sample is completely solid and it has moved into the solid-only field. Since the last drop of liquid is at the eutectic point E, its composition must be 60% Di and 40% An.

ITQ 7

The answers are illustrated in Figure 4.54.

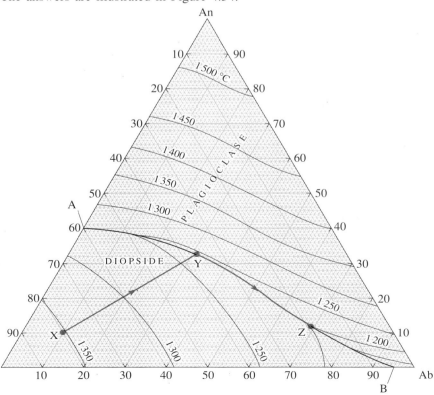

Figure 4.54 Answer to ITQ 7.

(a)   A composition of 80% Di, 10% Ab, and 10% An plots at point X in Figure 4.54. The liquidus surface is at $1350\,°C$ at this composition, and hence that is the temperature at which the first crystals start to form.

(b)   Because point X plots on the diopside-rich side of the cotectic curve (AB, Figure 4.54), diopside crystallizes first. With further cooling, the composition of the remaining melt moves directly away from the diopside apex of the triangle and intersects the cotectic curve at about $1245\,°C$ (point Y). At this temperature, plagioclase starts to crystallize with diopside. These two minerals crystallize together as the melt cools along the cotectic curve, until crystallization is complete (point Z).

(c)   The first plagioclase to crystallize at $1245\,°C$ (point Y) from this melt has the composition An_{80}. This is because the Ab:An ratio is the same in melts X and O of Figure 4.9, that is, $1:1$. The composition of plagioclase in the liquid is An_{50}, and from the plagioclase binary system (Figure 4.5) a liquid of composition An_{50} is in equilibrium with plagioclase crystals of composition An_{80}. The composition of the plagioclase changes from Y to Z (Figure 4.54) as the sample cools, and, since the crystals at Z have the same composition (An_{50}) as the plagioclase in the original liquid, then under equilibrium conditions they must be the last crystals of plagioclase to form.

(d)   The composition of Z (the last liquid to crystallize) in Figure 4.54 is the same as that of M in Figure 4.9 (An_{15}), because the composition of the plagioclase in the original sample is the same in each case. This can be determined from the plagioclase binary system (Figure 4.5). The composition of liquid in equilibrium with a plagioclase of composition An_{50} is An_{15}.

ITQ 8

(a)   In a simple binary eutectic system such as Di–An, the compositions of the minerals are fixed. Thus, whether or not they are in equilibrium with the liquid, their composition cannot change and the composition of the liquid at any particular temperature will therefore be the same whether crystallization takes place under equilibrium or non-equilibrium conditions.

(b)   As illustrated in Figure 4.7, the most highly fractionated liquids produced by non-equilibrium crystallization in the plagioclase system will be richer in albite than those produced under equilibrium conditions.

ITQ 9

The answer is illustrated in Figure 4.55.

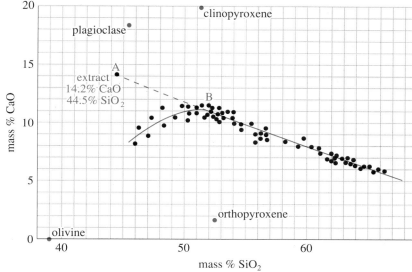

Figure 4.55 Answer to ITQ 9.

The CaO (14.2%) and SiO_2 (44.5%) contents of the aggregate phenocryst assemblage plot at point A, on-line with the trend of the lavas with 52–60% SiO_2. Removal of an extract of that composition would therefore drive the composition of the residual liquid along the trend of those lavas. We may conclude that fractional crystallization of that phenocryst assemblage could be responsible for the variations in CaO and SiO_2 in the lavas with 52–60% SiO_2.

ITQ 10

Between 53 and 55% silica, the Al_2O_3 content rises steeply, suggesting the crystallization of a phase containing little Al_2O_3, such as diopside. In this range, MgO decreases sharply confirming segregation of clinopyroxene. Hence the path A–B (Figure 4.14) describes the change in liquid composition due to fractional crystallization of clinopyroxene.

Between 55 and 61% SiO_2, Al_2O_3 is virtually constant whereas MgO decreases less sharply than before. This suggests that a second mineral is fractionating from the magma, which contains significant Al_2O_3, but little or no MgO, such as plagioclase. This is consistent with fractional crystallization along the cotectic curve (B–C) in Figure 4.14.

ITQ 11

In general, the partition coefficients are much less than 1 in each case for both elements, so fractional crystallization of any combination of these minerals would result in the steady increase in both Y and Zr in the residual liquid. The only exception is K_D for Y in clinopyroxene ($K_D = 1$). Fractionation of this phase would result in no change to the concentration in the magma.

ITQ 12

It is impossible to form a magma with $SiO_2 > 68\%$ (the Di–Ab eutectic composition) by fractional crystallization *within the ternary system Di–An–Ab*. However, basaltic rocks contain more elements than Si, Ca, Mg, Na, Al and O, and more minerals than diopside and plagioclase. Indeed, as crystallization proceeds, new minerals will form, and the ternary plot of Figure 4.9 will become less applicable to natural magmas. Other ternary systems must be introduced, which include quartz and alkali feldspar.

There is therefore no theoretical reason why granites of $SiO_2 = 74\%$ should not be formed from sufficient fractional crystallization.

ITQ 13

The answers are illustrated in Figure 4.56.

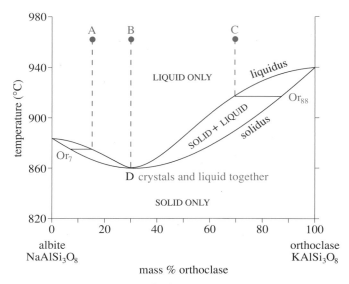

Figure 4.56   Answer to ITQ 13.

(a)   Melt A has a composition Or_{15}. When this cools to intersect the liquidus, the solid phase that forms is on the solidus at the same temperature and must have the composition Or_7. Melt B has the composition Or_{30}, and when this cools to intersect the liquidus, because the liquidus and solidus coincide, the crystals will have the same composition as the liquid Or_{30}. Melt C has a composition of Or_{70}. The crystals formed when this melt is cooled to the liquidus have a composition of Or_{88}.

(b)   If equilibrium between the crystals and the liquid is maintained throughout crystallization, then the composition of the last crystals to form will be the same as that of the total sample, that is: for A, Or_{15}; for B, Or_{30}; and for C, Or_{70}.

ITQ 14

The crystallization path is illustrated in Figure 4.57.

On cooling a liquid with composition R, crystals of pure quartz begin to form when the temperature reaches the liquidus (840 °C). With further cooling, the melt composition moves towards the cotectic curve AMB. The point at which the liquid will reach the cotectic curve can be determined by drawing a straight line from the quartz apex, through point R, to the cotectic curve. At this point (S in Figure 4.57), the melt will begin to crystallize alkali feldspar, which will be homogeneous and very rich in the albite component.

As both alkali feldspar and quartz continue to crystallize, the remaining liquid moves down the cotectic curve (towards M). As this occurs, the composition of the alkali feldspar changes by reaction with the melt — both crystals and liquid become richer in orthoclase.

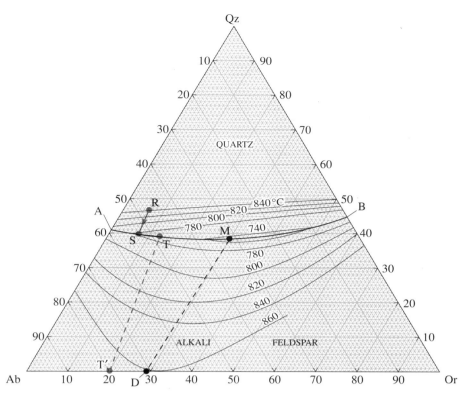

Figure 4.57 Answer to ITQ 14.

(a) If the equilibrium between crystals and liquid is maintained throughout crystallization, solidification is complete when the crystals of alkali feldspar have the same composition as the feldspars in the original samples (that is, approximately Or_{10}). From Figure 4.23, crystals of Or_{10} are in equilibrium with a liquid of composition Or_{20} (T' in Figure 4.57). The final liquid will lie at composition T, where the line Qz–T' intersects curve AMB.

(b) If equilibrium is not maintained, the remaining *liquid* moves further down the cotectic curve towards the minimum point M. If the crystals and liquid are separated, that is, if fractional crystallization takes place, then the final liquid will form a solid with virtually the same composition as that at the granite minimum (M).

ITQ 15

C_o and C_1 represent trace-element concentrations in the parental andesite and evolved granite, respectively.

(i) For Rb, $C_1/C_o = 140/70 = 2$.

From the curve for $D = 0.1$ on Figure 4.27, $F = 0.45$ for $C_1/C_o = 2$. In other words, 45% of the magma must be left to form a granitic liquid.

(ii) For Sr, $C_1/C_o = 100/340 = 0.29$.

It is not easy to obtain a precise value of F graphically for $D = 3.5$, but $F \sim 0.6$ from the curve for $D = 4$.

The results could be determined accurately from Equation 4.4, which would yield values of $F = 0.46$ and $F = 0.61$, respectively.

ITQ 16

(i) For Rb, $C_1/C_o = 2$.

From the curve for $D = 0.1$ on Figure 4.28, $F = 0.43$.

(ii) For Sr, $C_1/C_o = 0.29$.

For $D = 3.5$, F is very small.

Again, we can calculate these results more precisely, this time from Equation 4.3. This gives us $F = 0.44$ and 0.04 for Rb and Sr, respectively.

ITQ 17

(a) ^{87}Rb decays to ^{87}Sr; therefore (i) the ^{87}Rb/^{85}Rb ratio will decrease, (ii) the ^{87}Sr/^{86}Sr ratio will increase, and (iii) the ^{86}Sr/^{84}Sr ratio will not change with the passage of time.

(b)   A rock with no Rb cannot generate any ^{87}Sr and thus its ^{87}Sr/^{86}Sr ratio will not change.

(c)   Whether Sr is initially present or not, ^{87}Rb will decay and thus the ^{87}Rb/^{85}Rb ratio will decrease.

ITQ 18

(a)   The slope of the biotite–plagioclase isochron is *less steep* than that of the rock isochron. The slope equals λt, and thus the steeper it is, the larger must be the time t. Therefore, the minerals must be younger than the rocks.

(b)   The slope of the biotite–plagioclase isochron is

$$\frac{0.804 - 0.736}{4 - 1} = 0.022\,67$$

but the slope $= \lambda t$, and $\lambda = 1.42 \times 10^{-11}\,\mathrm{a}^{-1}$. Therefore,

$$0.022\,67 = \lambda t$$

$$t = \frac{0.022\,67}{1.42 \times 10^{-11}\,\mathrm{a}^{-1}}$$

$$\approx 1\,600\,\mathrm{Ma}.$$

The mineral isochron corresponds to an age of about $1\,600\,\mathrm{Ma}$. (The uncertainty on this age is $\pm 100\,\mathrm{Ma}$.)

ITQ 19

The answer is illustrated in Figure 4.58.

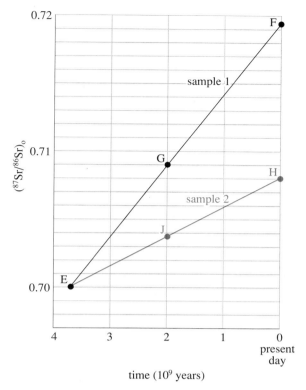

Figure 4.58 Answer to ITQ 19.

The second sample of the west Greenland granites has a present-day $^{87}Sr/^{86}Sr$ ratio of 0.708 0. 3 700 Ma ago, it had an Sr-isotope ratio of 0.700 1 assuming it formed from the same source as Sample 1. It therefore evolved along the line EH on Figure 4.58. 2 000 Ma ago it was at point J and had an $^{87}Sr/^{86}Sr$ ratio of 0.703 8.

ITQ 20

The answer is plotted on Figure 4.59

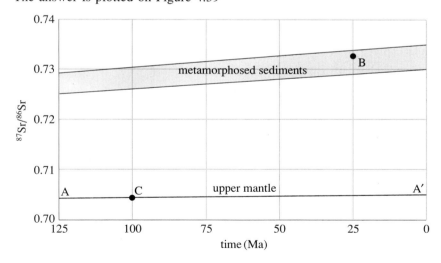

Figure 4.59 Answer to ITQ 20.

The Chilean granite plots at point C on the isotope-evolution diagram. It lies on the mantle evolution diagram and could therefore have formed by fractional crystallization of a melt derived from the mantle.

The Bolivian granite had an initial Sr-isotope ratio of 0.732 5, 25 Ma ago. If you plot this point on Figure 4.35 (B on Figure 4.59), it falls within the band for metamorphosed sediments, suggesting that this granite (and by implication the tin) was derived by remelting of the metamorphosed 300 Ma sediments.

ITQ 21

There appear to be no significant differences between high-silica and low-silica magmas between 200 and 50 Ma. This suggests that more silica-rich magmas have the same source as basaltic magmas. In other words, all magmas have formed from partial melting of the mantle and none have formed from crustal sources. Remember that fractional crystallization of a melt will not change its Sr-isotopic ratio, but if more silicic magmas result by contamination of mantle melts with crustal melts, then the resulting Sr-isotope ratios will be displaced above the ratios of basaltic rocks in a systematic way.

ITQ 22

For young magmas (less than 50 Ma old), there is a clear increase in initial Sr isotope ratios, many of which exceed 0.706 0, the maximum value ascribed to IAB in Figure 4.34. This is evidence that they have been derived from crustal melting or at least have been contaminated by crustal rocks.

ITQ 23

For each phase boundary, the polymorph on the high-pressure side has the highest density. Thus, the order of decreasing density is kyanite, sillimanite, andalusite. (In fact, the densities are kyanite, $3\,600\,\mathrm{kg\,m^{-3}}$; sillimanite, $3\,250\,\mathrm{kg\,m^{-3}}$; andalusite, $3\,150\,\mathrm{kg\,m^{-3}}$.)

ITQ 24

From Figure 4.42, the minerals wollastonite, calcite and quartz must lie on the boundary curve of the decarbonation reaction. This crosses the boundary between kyanite and sillimanite at a pressure of about $1.0\,\mathrm{GN\,m^{-2}}$ (Figure 4.40), which represents a minimum pressure for the assemblage.

ITQ 25

The initial Sr-isotope ratios for some of the younger samples from the batholith are larger ratios than the ratio on the mantle evolution line (Figure 4.45), but are much smaller than ratios within the band obtained from the sediments of southern Tibet. Therefore, although some of the melts may be simple fractional crystallization products of a mantle-derived gabbro (b), the younger melts with larger initial Sr-isotope ratios indicate contamination by crustal material (c).

ITQ 26

The high initial Sr-isotope ratio strongly suggests a crustal source. The restricted, silica-rich compositions of the granites (SiO_2 = 70–75%), equivalent to minimum melt compositions, support partial melting in the quartz–plagioclase–alkali feldspar system (Section 4.3.2). From Section 4.3.7, we know that crustal melting can occur in thickened crust due to the high content of heat-producing elements in crustal rocks. Since the intrusion age of 20 Ma is at least 20 Ma after initial collision, it seems likely that the granites have formed after the thickened sedimentary pile has heated up due to the high concentrations of heat-producing elements in crustal rocks.

ITQ 27

(a) (i) From Figure 4.46, geotherm A does not intersect the melting reaction and so no melting occurs. (ii) Geotherm B crosses the melting reaction (Equation 4.16) at depths of about 24 km, and this is where melting will occur. You should recall from Section 4.2 that no melting is possible where the geotherm crosses the wet granite solidus (at about 20 km) because of the absence of H_2O and alkali feldspar in the sediments. (b) At 24 km, the geotherm lies within the sillimanite field, which is the polymorph that is observed in the Himalayan metasediments (Plate 4.4).

ITQ 28

From Equation 4.3, which describes trace-element behaviour during partial melting,

$$C_1 = \frac{C_o}{D + (1 - D)F} .$$

Now, $C_1 = 300$, $C_o = 150$, and $D = 0.44$. Therefore

$$300 = \frac{150}{0.44 + (1 - 0.44)F}$$

$$0.44 + 0.56F = 0.5$$

$$F = 0.11.$$

An approximate value of $F \sim 0.1$ could be read off Figure 4.28 for $C_1/C_o = 2$, $D = 0.44$.

ITQ 29

(a) Slope of the isochron on Figure 4.48 is given by $(0.7600 - 0.7500)/(30 - 0) = 3.3 \times 10^{-4}$. From Equation 4.10, slope $= \lambda t$. Therefore

$$t = \frac{3.3 \times 10^{-4}}{1.42 \times 10^{-11}\,a^{-1}} = 23\,\text{Ma}.$$

(b) The initial isotope ratio for the granite is 0.7500. When this value at 23 Ma is plotted on Figure 4.49, it lies within the mica schist field and well below the migmatite field. Therefore, the Sr-isotope data support the mica schist as a source for the granite and rule out the migmatites.

SAQ ANSWERS AND COMMENTS

SAQ 1

(a) In the Di–Ab–An system (Figure 4.9), the cotectic curve slopes down-temperature from the Di–An to the Di–Ab eutectic. Thus, fractional crystallization, which produces lower-temperature liquids, drives them towards the Di–Ab eutectic.

(b) Anorthite contains much more CaO than albite (see Table 4.2), and thus liquids driven towards the Di–Ab eutectic by fractional crystallization will be *poorer* in CaO than those generated by equilibrium crystallization.

(c) Anorthite and diopside both contain less SiO_2 than albite (Table 4.2). Thus, since the liquid at P (Figure 4.9) contains more anorthite and diopside than a liquid at the Di–Ab eutectic, it must also be poorer in SiO_2.

SAQ 2

The SiO_2 and Al_2O_3 contents of the volcanic rocks from Table 4.4 are plotted in Figure 4.60. Note that the various letters and lines related to them are for the answers to SAQs 2–4.

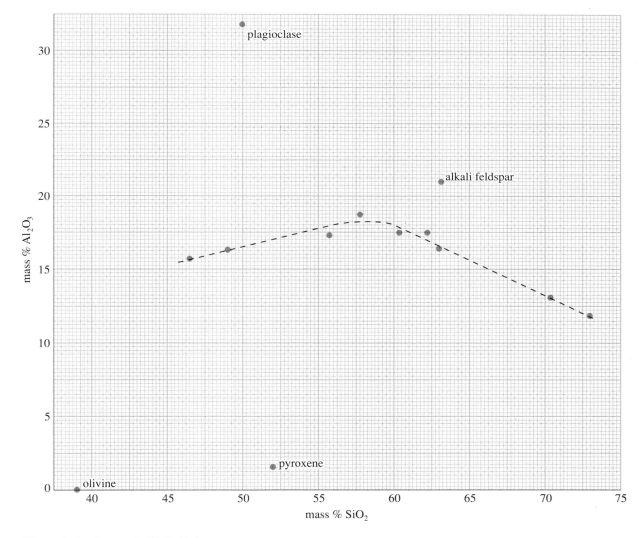

Figure 4.60 Answer to SAQs 2–4.

(a)   The trend of the samples of volcanic rocks in Figure 4.60 is not straight — rather, there is a marked change in slope (a kink) at 58–60% SiO_2 and at about 18% Al_2O_3.

(b)   A change of slope on such chemical variation diagrams signifies that a new mineral has started to crystallize, either on its own or with other minerals.

(c)   (i) From 47% to about 58–60% SiO_2, Al_2O_3 increases with increasing SiO_2. Assuming that this reflects fractional crystallization, the mineral extract must have less Al_2O_3 and SiO_2 than the liquid at 60% SiO_2.

(ii) From about 60 to 73% SiO_2, Al_2O_3 decreases with increasing SiO_2; thus the extract must have *more* Al_3O_3, but *less* SiO_2 than the liquid at 60% SiO_2.

SAQ 3

The SiO_2 and Al_2O_3 contents of the minerals in Table 4.5 are plotted in Figure 4.60. None of these minerals plots on the extension of the trend of the volcanic rocks with 47–56% SiO_2, and we may conclude that the crystallization and separation of any *one* mineral could *not* by itself have controlled the chemical evolution of these rocks.

SAQ 4

(a)   In the rocks with less than 58% SiO_2, Al_2O_3 increases as SiO_2 increases.

(b)   The kink (at 58–60% SiO_2) in the trend of the analyses you plotted on Figure 4.60 signifies a change in the minerals that are controlling the evolution of the liquids. For liquids above 60% SiO_2, Al_2O_3 *decreases* as SiO_2 increases, and thus the extract must contain *more* Al_2O_3 than the liquid.

(c)   The feldspars (plagioclase and alkali feldspar) both contain more Al_2O_3 than the liquids, and crystallization of feldspar must become increasingly important in these high-SiO_2 rocks.

SAQ 5

(a)   For basalt A, Zr/Y = 2, and for basalt B, Zr/Y = 3. From Figure 4.18, basalt A lies within the IAB field and basalt B within the MORB field. A is therefore derived from an island arc and B from a mid-ocean ridge.

(b)   The IAB (basalt A) would have higher Rb concentrations because elements of large ionic size and low valency such as Rb are enriched in the mantle wedge beneath island arcs by fluids derived from the subducted slab (Figure 4.17).

SAQ 6

In the discussion of Figure 4.22, we suggested that two major problems with the fractional crystallization model were that (a) 60–70% fractional crystallization was probably necessary and the resulting large quantity of crystals has not been found, and (b) basaltic liquid probably represents 20% partial melting of peridotite and thus unrealistically large volumes of mantle are required to produce reasonable quantities of granitic magma.

SAQ 7

The answers are illustrated in Figure 4.61.

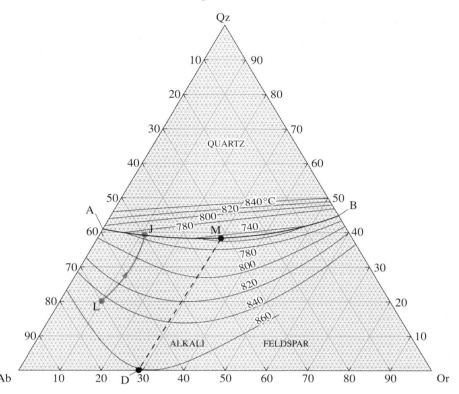

Figure 4.61 Answer to SAQ 7.

(a) The sample consisting of 20% quartz, 70% albite and 10% orthoclase plots in the alkali feldspar field at point L. Therefore the first mineral to crystallize is alkali feldspar.

(b) This sample plots on the albite-rich side of the feldspar minimum (the line DM). The phase diagram for the pure Ab–Or system (Figure 4.23) illustrates that the first crystals have more albite and thus *less* orthoclase than the coexisting liquid.

(c) The first crystals appear at about 845 °C.

(d) If equilibrium is maintained throughout crystallization, the feldspar becomes richer in orthoclase as the temperature falls (see Figure 4.23).

(e) Crystallization of the alkali feldspar, which will be richer in albite than the liquid, drives the liquid towards the cotectic along a path that is curved because the crystallizing feldspar is becoming richer in orthoclase. At the cotectic, the first crystals of quartz appear at a temperature of about 775 °C (at J).

(f) No, because crystallization will cease when the composition of the feldspar crystals is the same as those in the original sample, $Or_{12.5}$. The coexisting liquid will have an approximate composition of Or_{20} (Figure 4.23), which is more albite-rich than the minimum melt composition.

SAQ 8

The results from Table 4.8 are plotted in Figure 4.62.

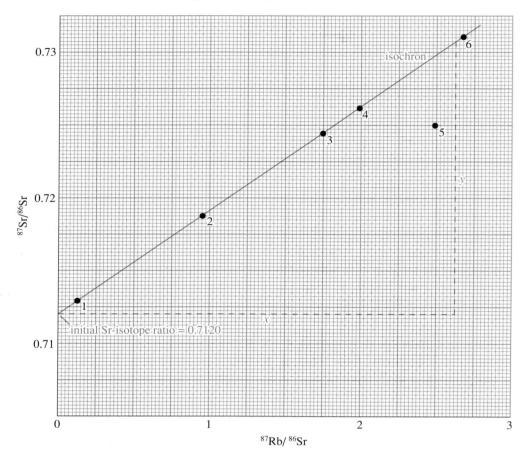

Figure 4.62   Answer to SAQ 8.

Five of the six points lie on a straight line; sample 5 does not. The slope of the straight line (the isochron) equals y/x.

$$y = 0.7305 - 0.7120 = 0.0185$$

$$x = 2.60 - 0 = 2.60$$

Therefore

$$\text{Slope} = y/x = \frac{0.0185}{2.60} = 0.00712.$$

But the slope $= \lambda t$, and λ (the decay constant for ^{87}Rb) $= 1.42 \times 10^{-11} \, a^{-1}$. Thus

$$0.00712 = 1.42 \times 10^{-11} \, a^{-1} \times t$$

$$t = 501 \times 10^{6} \text{ years}$$

The slope of the isochron in Figure 4.62 corresponds to an age of 501 Ma.

SAQ 9

There are three possible reasons why sample 5 plots below the isochron in Figure 4.62:

(i)   its age is different from that of the other samples in Table 4.8;

(ii)   it had a different initial Sr-isotope ratio;

(iii) during the last 501 Ma, the Rb/Sr ratio has changed by some process other than that of radioactive decay, for example, chemical alteration during weathering.

Analytical error is not a reason, since the sample lies much further from the isochron than is accounted for by the stated uncertainties in their measured isotopic ratios.

SAQ 10

The initial Sr-isotope ratio of the isochron in Figure 4.62 is 0.712 0. If you plot that on Figure 4.32 at an age of 501 Ma, it falls below the evolution line for the granites. At the same time, the initial ratio lies well above the Earth evolution line at 501 Ma (Figure 4.34). Hence a crustal source is indicated, but a younger source than the metamorphosed granites, or one with a lower Rb/Sr ratio.

SAQ 11

(a)   The initial Sr-isotope ratios lie close to the Earth evolution curve (Figure 4.34) and hence to the upper mantle evolution curve. Therefore, the granites must have formed from the mantle and represent evidence for crustal growth 500 Ma ago.

(b)   The Sr-isotope data only provide information on source regions at the time the magma was formed. Consequently, we can say nothing about crustal growth today from the data provided.

SAQ 12

(a)   Aragonite is stable at higher pressures and therefore is the denser polymorph of $CaCO_3$.

(b)   Calcite is stable at high temperatures (for a given pressure) and therefore will have a more disordered structure.

SAQ 13

Before collision, some crustal growth occurs at the active continental margin (i.e. magmas are derived from a mantle source) as can be seen from the presence of gabbros in the trans-Himalayan batholith and from the magmas of low initial Sr-isotope ratios of Figure 4.45. After collision, the only magmatism results entirely from crustal melting (ITQ 26). Collision therefore results in a decreased crustal growth rate at that particular plate boundary.

SUGGESTIONS FOR FURTHER READING

Wilson, M. (1989) *Igneous Petrogenesis*, Unwin Hyman, 466 pp. For students wishing to look further into the relationship between plate tectonics and igneous geochemistry. This book is particularly informative on volcanic rocks and their origins.

Best, M. G. (1982) *Igneous and Metamorphic Petrology*, Freeman & Co., 630 pp. A comprehensive and well-illustrated account of the basic principles that form the basis of igneous and metamorphic petrology.

Clarke, D. B. (1992) *Granitoid rocks*, Chapman & Hall, 283 pp. The complete guide to granites (geochemistry, mineralogy, field relations, and origin).

Cox, K. G., Bell, J. D. and Pankhurst, R. J. (1979) *The Interpretation of Igneous Rocks*. Allen & Unwin, 450 pp. Strictly for students interested in developing their understanding of quantitative geochemistry.

S339 Block 6 *Understanding the Continents: Tectonic and Thermal Processes of the Lithosphere. The Himalayas.* The Open University, 59 pp. The only available comprehensive account of magma formation in collision zones.

ACKNOWLEDGEMENTS

I thank Stuart Scott, Ray Macdonald, Nick Rogers, Simon Kelley, Colin Hayes, David Darbishire and the late Richard Thorpe for helpful comments on the manuscript during its evolution, and Eira Parker for typing and modifying the text more times than I care to remember.

Grateful acknowledgement is made to the following sources for permission to reproduce material in this block:

Figure 4.1: adapted from M. Wilson and J. P. Davidson (1984), 'The relative roles of crust and upper mantle', *Philosophical Transactions of the Royal Society of London*, A310, The Royal Society;   *Figure 4.2*: adapted from P. J. Wyllie (1982), 'Subduction products according to experimental prediction', *Bulletin of the Geological Society of America*, 93, © 1982 The Geological Society of America Inc.;   *Figure 4.3*: adapted from R. N. Anderson, S. E. Delong and W. M. Schwarz (1980), 'Dehydration, asthenospheric convection', *Journal of Geology*, 88, © 1980, The University of Chicago Press;   *Figure 4.6*: E. W. Heinrich (1965), *Microscopic Identification of Minerals*, McGraw–Hill Inc;   *Figure 4.20*: adapted from M. Wilson (1989), *Igneous Petrogenesis*, Unwin Hyman, currently published by Chapman and Hall;   *Figure 4.21*: M. A. Bussell and W. S. Pitcher (1985), 'The structural controls of batholith emplacement', in W. S Pitcher, M. P. Atherton, E. J. Cobbing and R. D Beckinsale (eds), *Magmatism at a Plate Edge*, © 1985 Blackie and Son Ltd;   *Figures 4.27 & 4.28*: adapted from K. G. Cox , J. D. Bell and R. J. Pankhurst (1979), *The Interpretation of Igneous Rocks*, Unwin Hyman, currently published by Chapman and Hall;   *Figure 4.33*: G. C. Brown , R. S. Thorpe and P. C. Webb (1984), 'The geochemical characteristics of granitoids', *Journal of the Geological Society*, Vol 141, © Geological Society 1984.

INDEX FOR BLOCK 4

(**bold** entries are to key terms; *italic* entries are to tables and figures)

BLOCK 4 COLOUR PLATE SECTION

Plate 4.1 The Aeolian Island Arc north of Sicily. In the foreground is the crater of the island of Vulcano. A chain of islands can be seen, the far one being Stromboli.

Plate 4.2 Andesitic cone of volcano from northern Chile, central Andes.

Plate 4.3 Blue blades of kyanite growing within a quartz lens in a mica schist (from Glen Esk, Scotland).

Plate 4.4 Lenses of fibrous white sillimanite growing within biotite gneiss (from the Langtang Valley, Nepal).

Plate 4.5 Crystals of tabular white andalusite growing within slate adjacent to the Skiddaw granite (Cumbria).

Plate 4.6 Biotite gneiss resulting from deformation of granite (from Sierra de Grados, Spain).

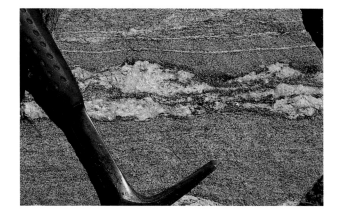

Plate 4.7 An example of migmatite. This is a rock which includes light coloured lenses of granitic composition within a dark biotite-rich gneiss (from Langtang Valley, Nepal).

Plate 4.8 Himalayan leucogranite sills intruded into darker metamorphosed sediments and migmatites (from Langtang Valley, Nepal).